AFAF AMARA-REKKAB

Goethite and Hematite for Zn (II) upgrading

AFAF AMARA-REKKAB

Goethite and Hematite for Zn (II) upgrading

Hydrometallurgical processes

ScienciaScripts

Imprint
Any brand names and product names mentioned in this book are subject to trademark, brand or patent protection and are trademarks or registered trademarks of their respective holders. The use of brand names, product names, common names, trade names, product descriptions etc. even without a particular marking in this work is in no way to be construed to mean that such names may be regarded as unrestricted in respect of trademark and brand protection legislation and could thus be used by anyone.

Cover image: www.ingimage.com

This book is a translation from the original published under ISBN 978-613-8-43538-9.

Publisher:
Sciencia Scripts
is a trademark of
Dodo Books Indian Ocean Ltd. and OmniScriptum S.R.L publishing group

120 High Road, East Finchley, London, N2 9ED, United Kingdom
Str. Armeneasca 28/1, office 1, Chisinau MD-2012, Republic of Moldova, Europe
Managing Directors: Ieva Konstantinova, Victoria Ursu
info@omniscriptum.com

Printed at: see last page
ISBN: 978-620-8-87286-1

Contents

Summary

Due to its characteristics and applications, such as galvanisation, cathodic protection and electronic circuits, zinc occupies an important place in the non-ferrous metals industry. It can be produced by pyrometallurgy or hydrometallurgy.

The hydrometallurgical zinc extraction process involves roasting the blende, followed by leaching to solubilise the zinc in the form of ZnSO4 using H2SO4 acid, which is then electroplated to produce purified solid zinc. The double acid and neutral leaching process (dissolution of zinc) in zinc refineries is not complete, achieving only a 90% extraction rate, resulting in solid metal waste in the form of ferrites that are insoluble in acid. Ferrites have the general formula $MeOPerO_3$, where Me is the metal element accompanying iron in zinc concentrates.

The insoluble residues from leaching and purification of the electrolytic solution are sent to landfill. Analyses carried out on the stored sludge showed a diverse composition of chemical elements in varying proportions, the most important of which are iron and zinc at 26% and 21% respectively, existing in several forms, particularly zinc ferrites. This is a source of pollution for the natural environment, particularly the marine coastline and groundwater.

The project in this thesis consists of the optimal extraction of zinc from leaching residues using acid extraction methods and iron precipitation methods in the form of a "Goethite" complex, an iron-rich compound that was then recovered as hematite, an iron pigment.

Acid extraction yielded zinc recoveries of 12 to 14% with a risk of parasitic dissolution of undesirable impurities, particularly iron. Zinc recovery by goethite precipitation is estimated at 19.43% with an extraction rate of 91.43%.

The final stage of our work consists of recovering the goethite from the second test and converting it into an iron oxide-based pigment, which is a product of great industrial and commercial value. Characterisation of the compound obtained by X-ray diffraction and Fourier transform infrared spectroscopy indicated that the hematite phase was the main phase of the compound with the total disappearance of goethite. The results of the pigment obtained are in agreement with several researchers.

Keywords: hydrometallurgy, residue treatment, zinc recovery, Goethite, Hematite.

General introduction

Zinc has been used since ancient times, but large-scale mining only began in Europe and America in the 19th century. Current global production is $4.9x10^{(6)}$ tonnes/year and reserves are estimated at $120x10^{(6)}$ tonnes. Zinc is used as an anti-corrosion agent (35-40% of production) and is also used in the composition of many alloys (brass), in the manufacture of batteries and paints.

Zinc is a chalcophile element, with a content of 80 ppm in the Earth's crust. It is mainly found in bedrock in the form of sphalerite (or blende) and wurtzite (ZnS), and to a lesser extent in zincite (ZnO), smithsonite (or calamite) ($ZnCO_{(3)}$), willemite (Zn2SiO4), zinkosite (ZnSO4), franklinite (ZnFe2O4) and hopite ($Zn_{(3)}$ (PO4)24$H_{(2)}$ o).

Zinc extraction is carried out using two methods (hydrometallurgy and pyrometallurgy) from ores after processing and enrichment. The dry or pyrometallurgical process involves the production of zinc by heating at different temperatures in roasting furnaces and reduction furnaces. It is obtained at a purity of approximately 98% pure metal, with the main impurities being lead, cadmium and iron. The process of extracting zinc by hydrometallurgy involves, after roasting the blende, leaching operations aimed at solubilising the zinc in the form of ZnSO4 by dissolution in H2SO4 acid, which, through electrolytic deposition, will produce purified solid zinc.

In Algeria, zinc is produced by hydrometallurgy by ALZINC, a subsidiary of METANOF located in Ghazaouet, in the wilaya of Tlemcen, approximately 640 km west of the capital Algiers. The plant was designed to process ore from El ABED and currently processes ore imported from Peru. According to the simple double leaching process, the rate is only 90%, but with new technologies for treating residues (jarosite, goethite, paragoethite, hematite), the extraction rate can now reach 96%. This is why ALZINC, which still uses the first process, is forced to increase the zinc extraction rate. To counteract the rising price of raw materials (miners are asking for more because producers are extracting more),

The double acid and neutral leaching process (zinc dissolution) is not complete, resulting in solid metal discharges in the form of ferrites that are insoluble in acid. Ferrites have the general formula MeOГerO3, where Me is the metal element accompanying iron in zinc concentrates.

In 2006, the leaching residues were transported and stored in a technical landfill site. Current environmental regulations stipulate that this type of waste can no longer be landfilled and must be recovered. Analyses carried out on the stored sludge showed a diverse composition of chemical elements in varying proportions, the most important of which are iron and zinc, existing in several forms, particularly zinc ferrites. This is a source of pollution for the natural environment, particularly the marine coastline and groundwater.

This work concerns a process for recovering certain elements, mainly metals and in particular zinc, from zinc-bearing material leaching residues.

Through this study, we will propose a technical solution for recycling zinc from leaching residues using a process known as "Goethite". The aim is to find a use for this residue by transforming it in a simple and economical way into an iron oxide-based pigment , which is a product of great industrial and commercial value with applications in many fields such as the paint industry.

This manuscript consists of four parts:

- The first part is a bibliographic summary that discusses zinc as a metal, its natural state, its properties and applications, and zinc extraction processes.

> The second part is devoted to a description of the ALZINC company, its various industrial discharges, waste management at the zinc electrolysis plant and their impact on the environment and human health.

> The third part focuses on the analysis techniques used and preliminary studies. The leaching residues produced by the plant have been characterised in detail using a number of analytical techniques: gravimetric, volumetric, atomic absorption spectroscopy, and chemical and mineralogical characterisation (atomic absorption spectroscopy, X-ray diffraction, inductively coupled plasma emission spectroscopy and Fourier transform infrared spectroscopy) of the treated leaching residues.

> The data collected and their interpretations are presented in the fourth part of the thesis.

Finally, we will conclude our work with a summary, in which we will outline recommendations for improving zinc extraction from solid metal leaching residues.

CHAPTER 1

Bibliographical references on zinc and extraction processes

Introduction

Zinc is a transition element with atomic number Z=30 and belongs to group IIB of the periodic table of elements. It ranks fourth after iron, aluminium and copper.
It is a white, slightly bluish, shiny metal. It breaks when cold, but becomes malleable and ductile between 100 and 150°C. In a humid environment, it becomes covered with a thin layer of hydrocarbonate, which protects it from oxidation.
It accounts for 0.012% of the Earth's crust and is present in water, soil and air. It is also an essential and indispensable component of the human body. The average daily requirement for an individual is between 9 and 15 mg, depending on whether they are a child or an adult.
Zinc is found in nature, usually in combination with other metals such as copper and lead. Zinc deposits are widely distributed around the globe. In those where it occurs in large quantities, the proportion of impurities in the rock is often high. The main ore is blende (or sphalerite) ZnS, which contains impurities such as lead, copper, manganese, calcium, cobalt, silver, etc.
Mines are located in Canada (British Columbia), the United States (Utah, Colorado, Idaho), Peru (Cerro de Pasco) and Australia. After extraction, the ore undergoes enrichment treatments consisting of particle size classification and an increase in the zinc content, which can reach up to 53%.
Zinc extraction is carried out using two methods (hydrometallurgy and pyrometallurgy) from the enriched ore.

- The dry process or pyrometallurgy involves the production of zinc by thermal means, heating at different temperatures in roasting furnaces and reduction furnaces, yielding zinc with a purity of approximately 98% pure metal. The main impurities are lead, cadmium and iron.
- The wet process, or hydrometallurgy, involves producing zinc by electrolytic deposition from a $ZnSO_4$ solution after leaching ZnO and purifying the zinc sulphate solution.

Zinc extraction using these two methods is accompanied by a number of environmental problems, including the accumulation of huge quantities of solid waste (metal compounds) and potential air pollution from dust, gases such as sulphur dioxide from the roasting of metal sulphide concentrates, and liquid waste (acid waste, wastewater rich in metal elements).
Zinc is characterised by its solubility properties for the formation of metal alloys (zamak, brass, etc.), its chemical properties for the formation of zinc salts used in several industrial fields (pharmaceutical industries), its protective electrochemical properties and its low mechanical resistance, zinc is an excellent element for its resistance to atmospheric corrosion (galvanisation and sacrificial anodes) and its mechanical properties, where, thanks to its ductility, rolled products, for building coverings, profiles for accessories used in construction (gutters and pipes for rainwater drainage systems), which led to its development in the 19th century.
In this chapter, we will present a study on zinc, its natural and anthropogenic state, its structure and properties, its different extraction methods, the various applications of zinc and its toxicity, the processes for treating residues, and we will conclude with a discussion on *the*

synthesis of scientific work on the recovery of solid waste.

1.1. Study on zinc

In this section, we will present the different natural forms of zinc, its structural state and its properties.

1.1.1. Natural sources

Zinc is found in varying quantities in many minerals and rocks. The most common ore in the lithosphere and the most widely used for metal production is zinc sulphide, known as blende or sphalerite (ZnS), which has a cubic structure **[1]**. It contains metallic impurities such as iron (dark blende), cadmium and silver.Blende is associated with other sulphides, such as galena (lead sulphide: PbS) **[2]**. It can also be found in other sulphide forms, such as marmatite ((Zn, FeS), which contains up to 14% iron.

Zinc is the fourth most produced metal in the world after iron, aluminium and copper, with global production continuing to rise. Today, around 65% of the zinc produced worldwide comes from mined ores.

Zinc is a chalcophile element (with a strong affinity for sulphur), mainly found in bedrock in the form of sphalerite (or blende) and wurtzite (ZnS), and to a lesser extent in zincite (ZnO), smithsonite (or calamite) (ZnCOs), willemite (Zn2SiÜ4), zinkosite (ZnSÜ4), franklinite (ZnFe2Ü4) and hopite (Zn3 (PÜ4)24H2Ü) **[2,3]**.

1.1.2. Geochemical behaviour

1.1.2.1. Geochemical behaviour in soils and sediments

The chemical forms of zinc in soils depend mainly on the composition of the parent rock. Like many other metal cations, zinc is more mobile in acidic environments. In addition to physical and chemical conditions, the mobility of zinc in soils is also governed by the proportion of clay, iron and manganese oxides and hydroxides, and organic matter (humic acids) present in the environment, as zinc is easily adsorbed or complexed in these phases.

In sediments, zinc is mainly found in the form of sulphides, due to the anoxic conditions prevailing in sediments when they are submerged **[3-5]**.

1.1.2.2. Geochemical behaviour in natural waters

The nature of soluble zinc species in natural waters depends on pH (Fig. 1.1). Under oxidising conditions, the dominant species are $Zn^{(2+)}$,ZnSO4 (aq), ZnCO3 (aq) and $ZnOH^{+}$. Up to pH 7.7, $Zn^{(2+)}$ is very much in the majority. As the pH increases, the neutral species Zn (OH)2 becomes predominant.

In addition to pH, soil and sediment composition also plays an important role in the solubility of species present in interstitial water. In particular, in soils rich in organic matter, zinc can be complexed by soluble organic molecules (small molecules, fulvic acids) and thus represent a significant proportion of the zinc in solution **[3-5]**.

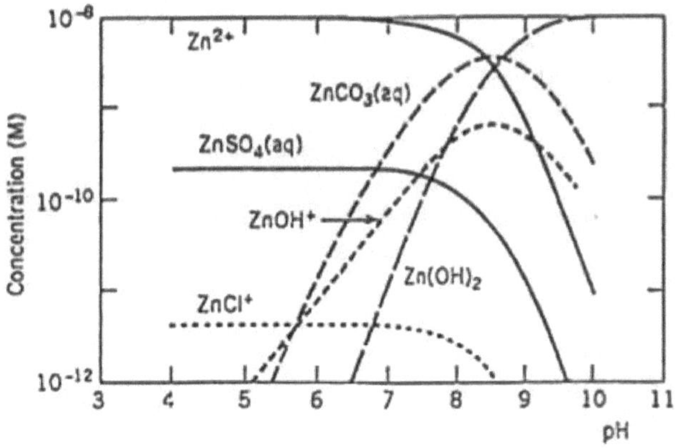

Figure 1.1: Main soluble inorganic zinc compounds in natural waters **[3]**

1.1.3. Anthropogenic sources

Human activities that have contributed to zinc inputs into the environment include manufacturing, iron and steel and non-ferrous metal industries, energy conversion (combustion of fossil fuels), waste incineration, road traffic (tyre wear and road surface wear), and manure spreading (zinc inputs in pig feed) [**6**].

1.1.4. Structural state

Zinc is one of the metals with a "compact hexagonal" stacking structure. Figure 1.2 shows the hexagonal lattice and the adjacent lattices required to represent the tetrahedral sites inside the lattice and the octahedral sites. The space group is p63/mmc. The lattice parameters are a=266 pm and c=495 pm [**7**].

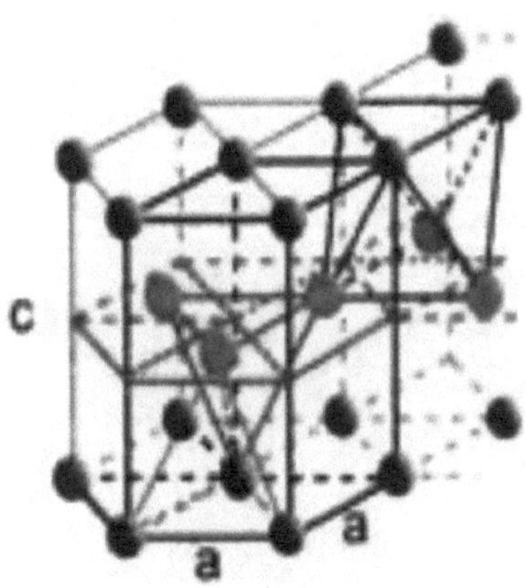

Figure 1.2: Schematic representation of the hexagonal lattice of zinc

1.1.5. Properties

Like all other metallic elements, zinc is characterised by a set of chemical and physical properties (melting point, resistivity , thermal and electrical conductivity, etc.) due to its delocalised metallic bond, and by its mechanical properties of plasticity. Due to the complete filling of its valence electron shells ($3d^{(10)}$ $4s^{(2)}$), zinc has only one oxidation state (+II) and there is no stabilisation of complexes conferred by the crystal field **[3].**

1.1.5.1. Chemical properties

Zinc is highly reactive in aqueous environments, as shown by the potential-pH diagram for the **Zn-H2O** system at **25°C** (Figure 1.3), which shows the immunity, corrosion and passivation zones. According to this diagram, zinc has no stability zone in common with water. It therefore oxidises with the release of dihydrogen in acid or strongly basic solutions that have been deaerated, in the form of divalent ions: **Zn**$(^{2+})$ in acidic environments and **ZnO2**$(^{2-})$ in strongly alkaline environments. At neutral or slightly basic **pH**, the oxidation of zinc leads to the formation of hydroxide. The introduction of carbonate ions into the system has the effect of broadening the passivation range towards slightly acidic **pH values**, following the formation of zinc hydroxycarbonate. The position of zinc in the standard electrode potential scale justifies its use as a sacrificial protective coating for carbon steel or low-alloy steel parts **[1]:**

$$E°(Zn^{2+}/Zn) = -\ 0{,}76\ V/ESH,\ E°(Fe^{2+}/Fe) = -\ 0{,}44\ V/ESH.$$

The behaviour of zinc coatings in natural atmospheres depends on the aggressiveness of the environment (rural, marine, urban or industrial) and in particular on the presence or absence of certain pollutants **[1].**

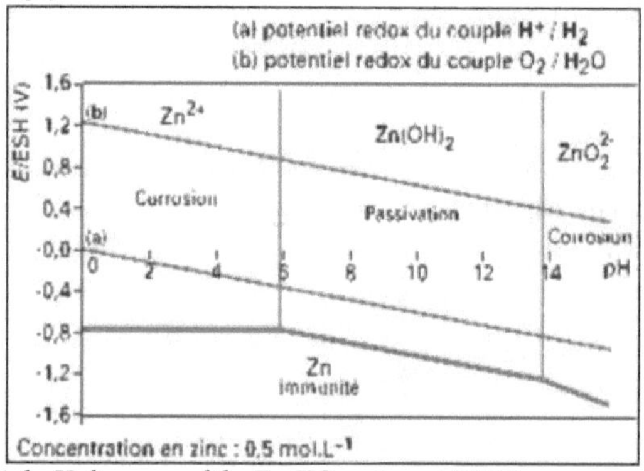

Figure 1.3: Potential-pH diagram of the Zn-HO system at 25 °C

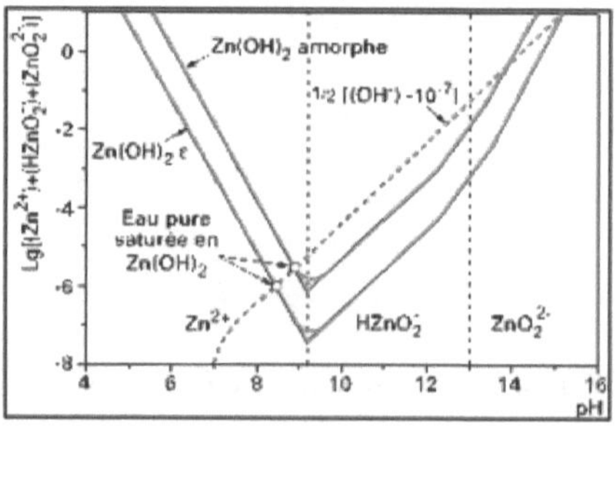

Figure 1.4: Influence of pH on the solubility of zinc hydroxides at 25°C

Metallic zinc and its oxide are amphoteric. Their solubility depends heavily on pH (Figure

1.4). Zinc forms compounds of varying solubility (sulphates, chlorides and carbonates) with many chemical agents, which play a key role in zinc corrosion. The zinc ion (Zn($^{2+)}$)has an ionic radius of 0.74 Å (Pauling scale) and a hydration number of 16. Its electronic configuration gives it the property of forming tetrahedral complexes in solution, mainly with dissolved ammonia and hydroxyl, halide, tartrate and cyanide ions **[1, 4,5].**

Zinc has the ability to form numerous alloys with other metals, both as a main constituent and as an additive (brass, zamak, nickel silver, bronze, etc.).

1.1.5.2. Physical properties

According to the periodic table of elements, zinc (Zn- Z= 30, molar mass: 65.38 g/mol) is the first element in column 12, which also includes cadmium (Cd), mercury (Hg) and ununbium (Uub). Zinc is a white, slightly bluish, shiny metal with a lamellar crystalline fracture. It is brittle at room temperature but malleable at 100 to 150 °C. Zinc has isotopes ranging from ($^{60)}$Zn to ($^{73)}$Zn. However, it has five stable isotopes: ($^{64)}$Zn (48.9%), ($^{66)}$Zn (27.8%), ($^{67)}$Zn (4.1%), ($^{68)}$Zn (18.6%), ($^{70)}$Zn (0.5%) and twenty-five other unstable isotopes and isomers. Some of the physical and chemical properties of this element are summarised in Table 1.**1 [1, 8, 9]:**

Table 1.1 - Main physical and chemical properties of zinc	**Physical and chemical properties of zinc**
Chemical symbol	Zn
Atomic number	30
Group, period, block	12, 4, d
Electron configuration	[Ar] $3d^{10}$ $4s^2$
Atomic mass	65.37 g · mol^{1}
Density (at 25 °C)	7,133 kg · m^{5}
Crystal structure	Hexagonal, c/a = 1.856
Electronegativity	1.65 (Pauling scale)
Oxidation states	0, π
Standard potential/ESH	- 0.76 V
Melting temperature	419 °C
Boiling point (760 mm Hg)	907 °C
Electrical resistivity (20 °C)	5.96 μΩ · cm
Thermal conductivity (solid 18 °C)	113 W · m^{1}· K^{1}
Atomic radius	1.35°A
Ionic radius	0.83°A

1.1.5.3. Mechanical properties

The mechanical properties of zinc are relatively low and depend heavily on its degree of purity. Thermomechanical treatments influence the final properties of zinc due to its easy recrystallisation in relation to its melting point. The tensile strength of zinc with a purity **of 99.995%** is low. Depending on the degree of work hardening, it can vary between **100** and **130 MPa.** For the same reasons, the hardness of zinc is between **30** and **35 Hv** . Zinc is malleable and can be rolled between **100** and **200 °C.** It becomes brittle at higher temperatures **[1, 4, 5].**

1.1.6. Areas of application

Steel protection is the main use of zinc. It accounts for approximately **40%** of global production, and to date there is no known substitute material that offers the same advantages: low melting point, good electrochemical properties and good atmospheric resistance **[10]**. Figure 1.5 shows a diagram of the different areas of application for zinc [**10-12]:**

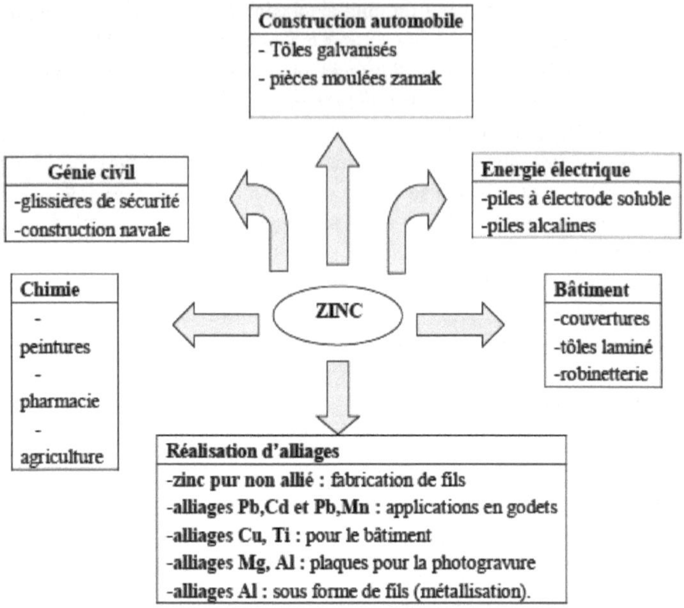

Figure 1.5: Various areas of application for zinc

1.1.7. Toxicity

Zinc is a trace element that is essential to all living species. It is a component of metalloproteins involved in many physiological reactions. In humans, zinc deficiency causes skin lesions and a weakened immune system. The recommended daily intake of zinc is between 5 and 40 mg, with toxic and lethal doses being 150-600 mg and 6 g respectively [**3, 13]**. Zinc is suspected of being carcinogenic in high doses.

In the workplace, zinc is absorbed through the respiratory tract (fumes, dust, vapours), with negligible skin penetration, although this is possible for soluble forms. Penetration through the digestive tract is mainly linked to dietary zinc (20-40% absorbed in the small intestine). In the blood, 85% of zinc is intraerythrocytic. All tissues and fluids contain zinc, mainly in muscles (63%) and bones (28%). Zinc induces metallothionein, to which it binds in the mucous cells of the intestinal tract. Elimination is mainly faecal (70%) and incidentally urinary (10%) and sudorific. The biological half-life is 200 to 500 days [**14]**.

1.2. Zinc extraction processes

Zinc is extracted using two methods: pyrometallurgy or the dry process, and hydrometallurgy or the wet process **[15]**. Pyrometallurgy was the first method used, dating back to ancient times. Hydrometallurgy, on the other hand, only emerged at the end of the 19th: zinc electrolysis was not carried out until 1916. Hydrometallurgy has developed significantly since the beginning of the century and has often overtaken pyrometallurgical processes for the production of many metals (Zn, Ni, Cu). Initially used for the treatment of ores or concentrates, it has been used for the treatment of metal waste for the past fifteen years.

Compared to pyrometallurgy, hydrometallurgy is less energy-intensive because the operations are carried out at much lower temperatures. Size is also an important factor, as small units can be designed at reduced cost. It also allows the processing of poorer ores while improving refining and extraction yields **[16].** Ore processing is necessary prior to these processes.

1.2.1. Ore processing (enrichment)

The purpose of enrichment is to separate a zinc concentrate from the gangue, as well as lead, copper and even pyrite concentrates. This is done physically. In the first stage, the ore must be ground sufficiently to release the individual metal-bearing particles from the waste rock. This release size is on average between **100** and **200 µm.**

In a second stage, the various useful minerals are separated by flotation, taking advantage of the relatively low surface wettability of sulphides: during pulping, using suitable conditioning, this wettability is selectively modified using organic and mineral additives (xanthate, pine oil, olein, zinc sulphate, copper sulphate, etc.) and by adjusting the **pH.** The metal-bearing particles of the same type are thus rendered hydrophobic one after the other.

The "washing plant" is installed directly at the mine. In most cases, the lead concentrate is floated first, followed by the zinc concentrate, but in rare cases it is more advantageous to separate a mixed concentrate first and then separate the sulphides from each other. Separations are never perfect and the zinc concentrate, due to the presence of mixed grains, always contains a small amount of lead and sterile material **[17].**

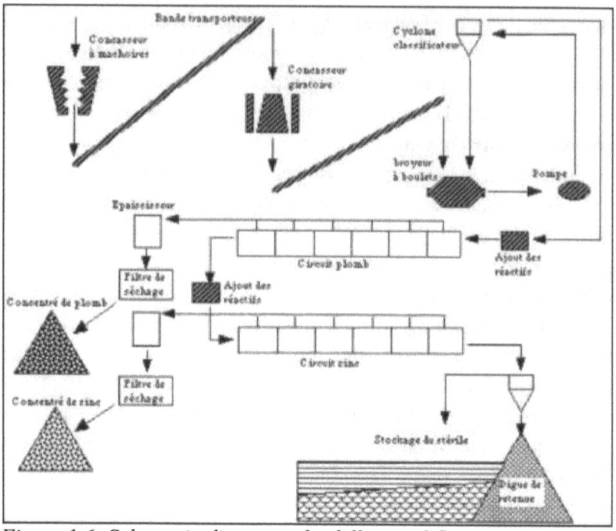

Figure 1.6: Schematic diagram of a differential flotation circuit **[15]**

1.2.2. Thermal extraction of zinc

The ore is converted into ZnO oxide, which is reduced by coal. The zinc sulphide is treated by roasting and reduction and the metal is volatilised (boiling point: 907°C) **[18].** The pyrometallurgical process involves four stages, including two stages of refining depending on the desired degree of purity.

1.2.2.1. Roasting of sulphides

Zinc metallurgy uses zinc oxide. To obtain this oxide, the concentrate is roasted at around 950°C, through the following exothermic reaction **[19]**:

$$2\,ZnS + 3\,O_2 \longrightarrow 2\,ZnO + SO_2 \quad (1.1)$$

Parasitic reactions occur at relatively low temperatures (500 to 600°C) and lead to zinc sulphate. However, zinc sulphate decomposes at temperatures above 900°C and its content is fairly low in ZnO calcine. In practice, the roasting temperature is generally between 910 and 980°C. The temperature is maintained by adjusting the feed rate of the concentrate into the fluidised bed, as the amount of heat produced is proportional to the feed rate. The residual sulphur content of the calcine is 1.5 to 2% in the form of uncalcined sulphide. The sulphur dioxide collected is used to manufacture sulphuric acid.

1.2.2.2. Oxide reduction

Zinc oxide is reduced by carbon monoxide, with the zinc obtained in a vapour state (Te = 907°C):

$$ZnO + CO \longrightarrow Zn_{gaz} + CO_2 \quad \Delta H = 50\ Kcal\ (1.2)$$

Carbon monoxide is produced from coal. The reduction of ZnO requires a high CO content supplied by the equilibrium:

$$CO_2 + C \longrightarrow 2\,CO \ (1.3)$$

Thanks to a sufficiently high temperature (over 950°C). Furthermore, it is necessary to cool the gases formed as quickly as possible in order to prevent the re-oxidation of zinc.
The thermal route is carried out in two processes **[18]**.

a) Reduction by carbothermy

It is carried out in a "vertical crucible furnace". The crucible is made of silicon carbide (SiC) brick. It is continuously fed from the top of the crucible with calcined carbon particles. Carbothermic heating is endothermic. Heating is carried out from the outside over the entire height of the crucible at a temperature of 1280-1320°C. The CO gas and metal vapours are carried to a condenser. The zinc is extracted from the gas mixture (CO+Zn) by condensation. Rapid cooling is required to prevent reoxidation due to air ingress. This is achieved by spraying liquid zinc **[20]**.

b) Reduction by CO in the presence of carbon

This process is used to treat mixed zinc and lead sulphide concentrates to obtain zinc and lead. It also allows the recovery of copper, silver, gold, arsenic, antimony and bismuth contained in these concentrates. A solid charge consisting of particles of roasted and agglomerated ore and coke flows from top to bottom of the blast furnace and the reducing gas flows countercurrently through this charge. The reducing gas is produced by burning the coke with preheated air injected through nozzles located at the bottom of the blast furnace.

The agglomerate charge consists of reducible oxides ZnO (36 to 43%) and PbO (17 to 22%), and oxides: Fe_2O_3 (8 to 13%), AfO? (7%), CaO (4%), SiO_2 (4%), which will form the slag.

Reduction produces gaseous zinc and liquid lead. As it flows downwards, the lead dissolves elements such as copper, silver, gold, arsenic, antimony and bismuth. At the same time, a slag containing the unreduced oxides is formed. It is cast and separated from the liquid lead by settling.

The gases leaving the reactor contain 8% Zn, 11% CO_2 and 25% CO. The zinc vapours are condensed by spraying liquid lead, which circulates in the condenser in a counter-current flow. It takes 400 tonnes of lead to condense 1 tonne of zinc. At the condenser outlet, the Zn-enriched lead is cooled to 450°C and separated by liquidation **[20]**.

1.2.2.3. Zinc refining

The zinc obtained by thermal means has a purity of approximately 98% pure metal (main impurities: lead, cadmium, iron).

a) Refining by liquation

The lead-laden zinc is melted in a deep-hearth reverberatory furnace, as zinc and lead are immiscible in the liquid state; they separate into two layers, the upper layer consisting of fairly pure zinc, which is the standard quality metal.

b) Refining by fractional distillation

This process consists of fractional distillation in a tray column and is based on the difference in boiling points: cadmium 767°C, zinc 907°C, lead 1740°C. This produces a metal that is 99.993% pure, and the cadmium, which has a high market value, is separated **[18]**.

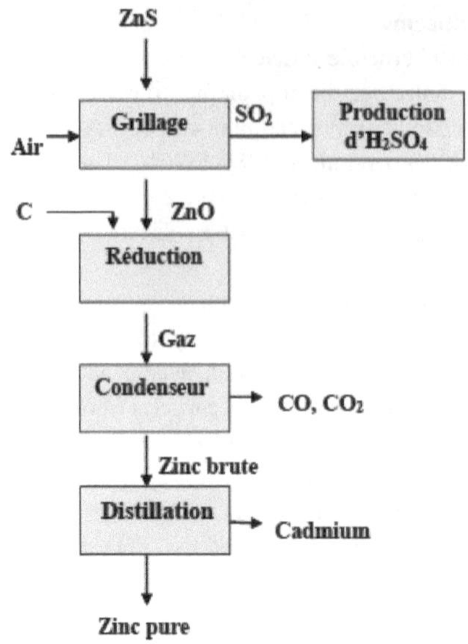

Figure 1.7: Pyrometallurgical process for zinc production **[21]**

1.2.3. Wet extraction of zinc

The preparation of zinc by electrolysis with insoluble anodes has developed, in particular, because the electrothermal process is much more delicate for zinc than for other metals and also because the electrolytic process allows the exploitation of complex zinc ore deposits whose thermal treatment would be particularly difficult **[22,23]**. Zinc hydrometallurgy takes place in four successive stages:

1.2.3.1. Roasting

In its initial form or enriched by flotation, the ore must first be converted into oxide by roasting. When heated to a sufficiently high temperature in the presence of air, zinc sulphide burns to form zinc oxide and sulphur dioxide. The latter is partially oxidised by the catalytic action of the iron oxide present to form sulphuric acid, which reacts with the zinc oxide to form zinc sulphate. The corresponding reactions are as follows for most common sulphide ores:

$$2\,ZnS + 3\,O_2 \rightarrow 2\,ZnO + 2\,SO_2 \quad (1.4)$$
$$2\,SO_2 + O_2 \rightarrow 2\,SO_3 \quad (1.5)$$
$$ZnO + SO_3 \rightarrow ZnSO_4 \quad (1.6)$$

The ratio of zinc oxide to zinc sulphate in the product obtained from roasting the ore depends on a number of factors (some of which are uncontrollable), such as the roasting temperature, the amount of air supplied to the mass during the operation, and the presence and activity of traces of other substances that can act as catalysts and therefore promote the formation of sulphuric anhydride. This partial sulphation of zinc oxide during roasting serves to compensate, to a certain extent, for the continuous and inevitable loss of sulphuric acid [**22**].

1.2.3.2. Leaching-hydrolysis

Leaching is an acid-base attack in which, in the presence of sulphuric acid and calcine, zinc passes into solution in the form of Zn^{2+} ions. The reaction:

$$ZnO(S) + 2\,H^{+} \rightarrow Zn^{2+}(aq) + H_2O \quad (1.7)$$

is exothermic and the solution is heated to 60°C.
In addition, this operation removes a first group of impurities: As, Ge, In. To do this, a small amount of $Fe^{2+}$is added, which is oxidised by air (1/5 oxygen) to Fe^{3+}. Then, by adding an excess of calcine, the pH is brought to 4.5-5 at around 50°C, which allows the removal of $Fe(^{3+)}$ and impurities in the form of sludge. Depending on the site, the solubilisation time varies from 1 to 4 hours. 75% to 90% of the zinc is recovered in the weakly acidic solution, analysis of which gives the following average percentages:

Table 1.2: Chemical composition of zinc sulphate solution after leaching

Elements	Zn g/l	Cu g/l	Cd g/l	Mn g/l	Mg g/l	Co mg/l	Ni mg/l	Ge mg/l
Title	150	0.1-1	0.1-0.5	2-10	2	1	2	0.01-0.4

The composition of the residue varies:
Zn from 1.5 to 25%; Fe from 10 to 30%; Pb from 2 to 10% **[24].**

1.2.3.3. Purification by cementation

The dissolved elements Cu, Cd, Co and Ni must also be removed from the solution. Metallic zinc is therefore capable of reducing Cu^{2+}, Cd^{2+}, $Co^{2+}$and Ni^{2+}ions to their metallic state according to:

$M^{2+} (aq) + Zn(s) \rightarrow Zn^{2+} (aq) + M(s)$ (1.8)

The purification process involves adding zinc powder with a particle size of between 20 and 40 µm to the solution. The metal forms, or cementation occurs, at between 40 and 90°C for 4 to 5 hours. This is followed by very effective liquid-solid separation by filtration through a fine cloth.
The zinc sulphate solution then contains: 0.2 mg/l of Cu^{2+}, 0.7 mg/l of Cd^{2+} and 0.8 mg/l of Co^{2+}. For other elements, the concentrations are even lower, ranging from IO^{-3} to 10^{-2} mg/l for germanium, antimony and arsenic **[24].**

1.2.3.4. Electrolysis

Electrolysis takes place between an aluminium cathode and a lead anode according to the reaction:

$ZnSO_4 + H_2O \rightarrow Zn + H_2SO_4 + ½ O_2$ (1.9)

The layer of alumina that forms on the cathodes prevents the zinc from adhering too strongly to the substrate, making it relatively easy to remove the zinc deposit. The cells consist of batteries of anodes and cathodes, with up to 86 cathodes covering an area of 1.6 m^2. Deposition occurs on both sides. The intensity per cell varies from 50,000 to 115,000 amps and production per cell from 1.25 to 3 tonnes/day. The deposition time before cathode extraction is 48 hours.
The medium is an acidic zinc sulphate solution with a composition of approximately 150 to 190 g/l of sulphuric acid and 50 to 70 g/l of zinc in the form of ZnSO4.
The temperature is maintained between 30 and 40°C. The basic reactions are:

• at the cathode: $Zn + 2e^- \rightarrow Zn \quad E° = 0.76\ V$ (1.10)

$2H^+ + 2e^- \rightarrow H_2$ (1.11)

• at the anode: $2H_2O \rightarrow O_2 + 4H^+ + 4e^- \quad E° = +1,23\ V$ (1.12)

The minimum electrolysis voltage is E = 2.022V. The overvoltages required to obtain the desired current density of 400 A/m^2 are 750 mV for the anodic reaction and 110 rnV for the cathodic reaction Zn^{2+} /Zn. The voltage drop due to the resistance of the electrolytic bath , for an inter-electrode distance of 30 mm and a current density of 400 A/m(2) is approximately 240 mV. The total voltage applied is therefore equal to 3.2 V. The cathodic current efficiency is around 90%, due to the release of hydrogen that occurs simultaneously with the deposition of zinc. The energy consumption per tonne of Zn produced is therefore around 2950 to 3500 kWh **[25].**

Return to ZnS electrolysis cell

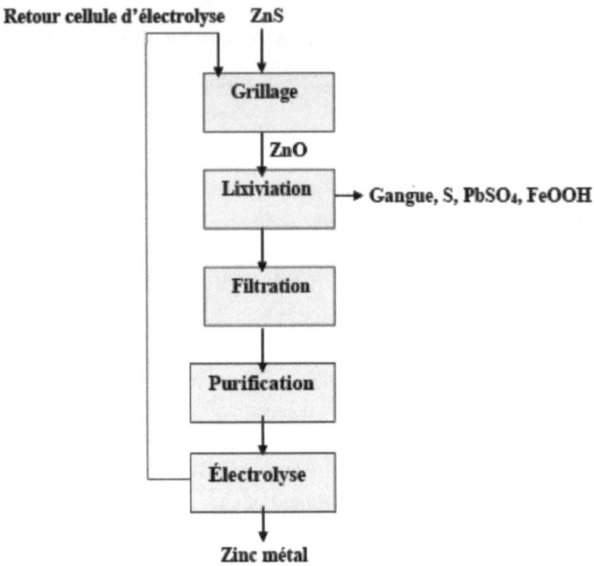

Figure 1.8: Hydrometallurgical process for zinc production **[21]**

1.3. Solid waste from zinc metallurgical operations

The zinc contained in ores is never entirely recovered in ingot form, as a significant percentage remains dispersed in metallurgical operations in the form of residues from various products during the different stages of extraction. Several processes **[17]** have been developed with the aim of increasing the dissolution rate in order to convert zinc into ionic form, on the one hand, and to recover zinc from solid waste, on the other. The double leaching process only achieves a solution rate of 90%, whereas with the Jarosite process, the rate can reach up to 97% **[26].**

1.3.1. Thermal processes

Historically, thermal processes were first used and are still used in about a third of plants. These may involve simple recycling in a lead plant or specific reduction treatment involving distillation of lead and zinc. The vapours, which burn at the outlet of the furnace, produce oxides that are easy to recover in a sulphuric environment. Two techniques can be mentioned in this context:

a) Waelz furnace: the dried residue is mixed with coke dust in a continuously operating rotary furnace, in which gas and solids circulate in countercurrent. Iron, copper and silver are found in a slag. Zinc is recovered from the gases in the form of an impure oxide, which is treated as calcine **[27].**

b) Cupola: the residue is agglomerated with pitch, mixed with coke and additives, limestone and silica, to form a fusible slag with part of the iron. The other part forms, with the sulphur in the charge, a sulphide, the matte, which is separated from the slag after melting by difference in density and which concentrates the copper and silver.

In both thermal processes, the cost price of these thermal processes has gradually increased (the cost of coke in particular), hence the renewed interest in the hydrometallurgical route, in which zinc is recovered through the simultaneous dissolution of iron. The subsequent removal of this element under favourable conditions has been the main difficulty to be overcome, which is why the different variants are named after the iron compound formed. In addition, it requires the use of suitable materials: rubber-lined or brick-lined tanks; agitators, pumps and heat exchangers made of stainless steel resistant to the highly acidic environment **[17].**

1.3.2. Wet processes

The hydrometallurgical treatment of zinc sulphide concentrates involves separating zinc from iron, as most zinc concentrates contain more than 5% iron. Several processes, such as hematite, goethite and jarosite, have been developed to remove iron from the solution prior to zinc electrolysis. In all three cases, iron is used to remove elements such as Al, Ga, ln, As and Ge, which are collected in the solution resulting from the attacks **[28-30].**

a) Jarosite process

The process of iron removal using jarosite was developed by the Asturiana zinc electrolysis company in Spain and by Det Norske Zinkkompari A/S in Norway in the early 1960s **[31].** Jarosites are basic iron sulphate complexes of the form 2 **M Fe3 (SO4)i(OH)6**, where M represents one of the alkali ions Na^+ , NH_4^+, K^+ . But perhaps also H_3O^+ or $\mathbf{Ag^+}$, $\mathbf{1/2Pb^{++}}$, etc. **[32].** The high concentration of $\mathbf{SO_4^{2-}}$ in the solution promotes this precipitation, which is enhanced by high temperatures and is more complete as we move from **Na** to K and **NH4.** The acidity must be further limited by adding calcine during precipitation, which corresponds to the reaction:

$$3Fe_2(SO_4)_3 + M_2SO_4 + 6ZnO + 6H_2O \longrightarrow M_2Fe_6(OH)_{12}(SO_4)_4 + 6ZnSO_4 \quad (1.13)$$

The ion **M** generally used is ammonium. The temperature is around **80-105°C** and atmospheric pressure at a pH = 1.5 **[33, 34].**

This precipitation causes a decrease in the sulphate content, which must be compensated for by adding sulphuric acid. It is most often carried out in two counter-current stages, the first ending at **10 g/l** of acid, the second at **pH** (5) to reduce the iron content of the solution as much as possible, which remains around 1 **to 2 g/l.** The pulp is decanted and the clear liquid is returned to the neutral solution. The solid phase is recycled to the acid stage, where it is separated and, before being stored at the landfill, washed with an acid solution to recover as much zinc as possible. In this regard, it is clear that the ferrites in the calcine used in this part of the circuit will not be dissolved, resulting in an inevitable loss of zinc of around one per cent. Elements such as **Al, Ge, In and As** precipitate at the most neutral stage and are only partially dissolved in the acid stage, hence their removal [17].

The Jarosite process offers an economic advantage in terms of investment and requires less labour. The main disadvantage of this process is the need to use expensive reagents. Furthermore, it is classified as hazardous waste and its disposal has become a major environmental concern worldwide **[35].** Jarosite results in a huge storage re s of residues compared to the zinc produced (50,000 tonnes per year of Jarosite for 100,000 tonnes per year of zinc) **[36].** Furthermore, it involves the consumption of calcine (and therefore a significant

loss of zinc) to neutralise the sulphuric acid formed during the reaction.
The recovery of jarosite residues has been the subject of extensive research.
V. Mymrin, A. Vazquez Vaamonde, Asokan Pappu, Mohini Saxena, and Shyam R. Asolekar **[37,38]** have described the possibility of using jarosite by adding 1–4% CaO or Portland cement as one of the components of new construction materials for various applications: as bases or sub-bases for roads, airfields and dams, instead of the crushed stone, gravel and sand currently used. They can also be used to make bricks, tiles and similar items.In addition to economic factors, it is very easy to use and no new waste is produced. Other authors, in particular Pappu Asokan, Mohini Saxena and Shyam R. Asolekar **[35]**, have proposed another way of recycling jarosite by mixing it with fly ash and soil clay for use as a building material. Another process has been proposed by W. Kunda and H. Veltman [**36**] for the transformation of jarosite by different thermal treatments into a hematite iron ore that can be used in the iron extraction industry.
* The following diagram shows the jarosite process:

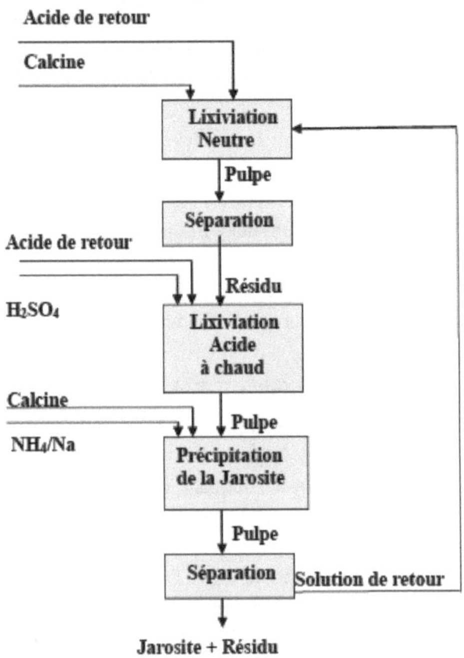

Figure 1.9: Jarosite process **[39, 40]**

c) Goethite process

The process for removing goethite iron was developed by the Société de Vieille Montagne in Belgium **[41]**. In this case, sulphate loss is avoided by precipitating the iron in the form **of FeO(OH).** The temperature range is **80 to 90°C.** The acidity must also be neutralised by adding calcine and the pH, measured at **70°C**, must be maintained above 2**.8** to **3.0.** As the iron-containing solution s out of the attacks at an acidity of **40 g/L,** it is first neutralised with calcine, and the solid residue is recycled to the attacks after separation. Given the low acidity during goethite precipitation and in order to avoid zinc losses due to excess calcine, some installations have been equipped with fairly sophisticated pH control systems. Furthermore, to remain within the goethite stability range, the **Fe($^{3+)}$**concentration must be low during precipitation **[42]**, hence two methods:

a) Continuous addition of the **Fe^{3+}**solution at the rate at which the iron is precipitated: this is the Electrolytique de Zinc process, developed by a company in Australia.

b) In a preliminary step, the ferric iron is reduced using blende: this is the Vieille Montagne process.

$$Fe_2(SO_4)_3 + ZnS \longrightarrow 2FeSO_4 + ZnSO_4 + S \quad (1.14)$$

By choosing the right conditions and a suitably reactive blende, this reaction takes place at 90°C with a good yield. The solid residue is recycled to the roasting stage. The ferrous iron is then reoxidised with air or, better still, with oxygen in the same tanks where the precipitation takes place:

$$2FeSO_4 + 2ZnO + {}^1/_2O_2 + H_2O \longrightarrow 2ZnSO_4 + 2FeO(OH) \quad (1.15)$$

Les agitateurs sont des turbines, pour assurer une bonne émulsion de l'oxygène.

The agitators are turbines, which ensure good oxygen emulsion. Elements in solution such as copper promote oxidation. It should be noted that the prior reduction of **Fe^{3+}** to **Fe^{2+}** reduces the amount of calcine required by **a third**, thereby reducing zinc losses in the form of compounds that are insoluble under these conditions by the same amount.

At the end of precipitation, the solution contains only about **1 g/L** of iron. It has been largely stripped of fluoride ions, which bind to goethite. The pulp is decanted, and the clear liquid is returned to the neutral solution. The solid phase is filtered, washed and stored **[17]**. During the goethite process, various precipitates may occur:

a -FeOOH Goethite, ß - FeOOH akagenite and hematite Fe_2O_3 with amorphous phases **[42,43]**.

The disadvantage of this process is the loss of a certain amount of zinc due to the use of calcine. The goethite process has the following advantages:

- Lower sulphuric acid consumption,
- No addition of alkali metals **[25, 44]**,
- Goethite can be used in all zinc extraction industries that use any type of acid (H2SO4, HCl, HNO3, etc.) as a leaching agent **[45]**,
- Goethite produces approximately 25 to 30% less waste than the conventional Jarosite-based process **[46]**,
- The reprocessing of leaching residues using the Goethite process has led to the development of new techniques for recovering lead and silver **[47]**.
- According to M.R.C. Ismael and J.M.R. Carvalho **[31]**, goethite offers some possibilities for additional treatment such as inerting and solidification. In both cases, an inert material could be used in the construction industry**.** Another possible solution was proposed by M. Romero and J.Ma. Rincon **[48,49]**, who demonstrated that goethite, when mixed with other raw materials, can be used in the formulation and production of ceramics and eyeglass lenses. At the same time, Jérôme Pradel, Simone Castillo, Jean-Pierre Traverse, Raoul Grezes-Besset and Michel Darcy **[50]** tested the high catalytic activity of goethite precipitate in coal hydro-liquefaction. However, another theory has been put forward by the Electrolytique du Zinc, Vieille Montagne company in Belgium **[51]** for the simple and economical conversion of goethite into an iron oxide-based pigment that can be used in the paint industry.

* The following diagram shows the goethite process:

Return acid

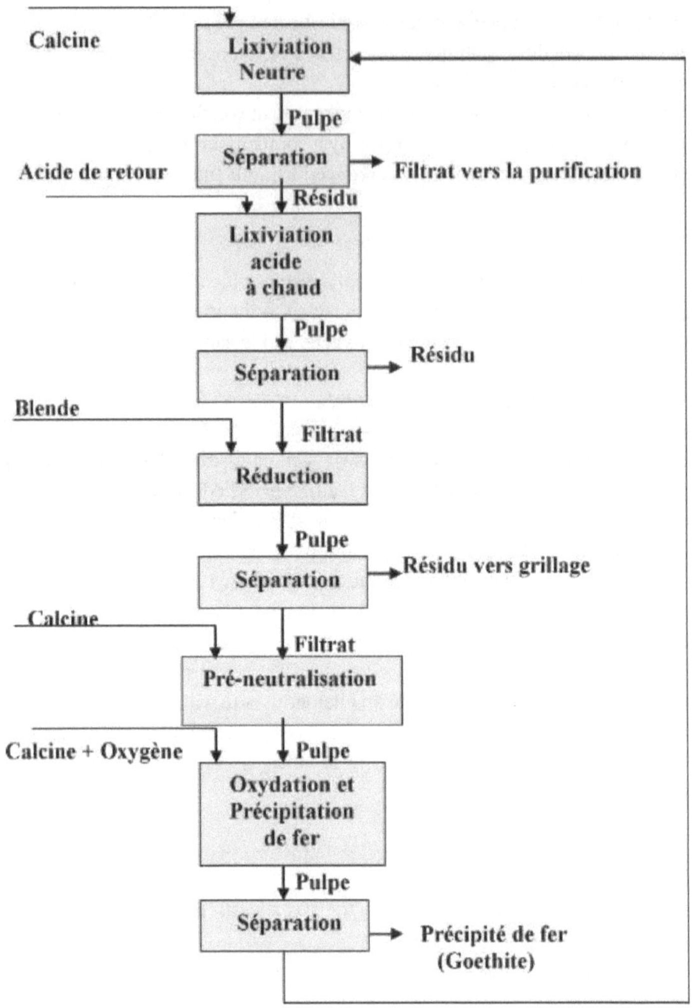

Filter

Figure 1.10: Goethite process **[52, 53]**

c) Hematite process

This process was first implemented in Japan in **1972** **[41].** Neutral leaching is unique to this process, as it takes place in a reducing environment in the presence of so2, which dissolves the iron into a divalent state, according to the reaction:

$$2Fe^{3+} + SO_4 + 2H_2O \longrightarrow 2Fe^{2+} + SO_4^{2-} + 4H^+ \quad (1.16)$$

Cu^{++} e is precipitated by H2S and the solution is neutralised to **pH 4.5** with lime, yielding marketable gypsum.

The iron is then precipitated at **200°C** under **18** atmospheres of oxygen **[17, 54]** according to the reaction:

$$2FeSO_4 + {}^1/_2O_2 + 2H_2O \longrightarrow Fe_2O_3 + 2H_2SO_4 \quad (1.17)$$

In this case, a sulphur residue and an iron residue in the form of hematite are obtained, which may contain between 0.2 and 1% zinc **[55]**. Hematite can be used in cement, in the iron industry or as a pigment **[31]**. However, this process has the disadvantage of involving high pressures, temperatures and sulphuric acid concentrations, requiring the use of a highly corrosion-resistant reaction chamber made from expensive materials such as titanium. Precipitation in the form of hematite would almost always require a considerable investment in the purchase of autoclaves. In fact, there are only two plants in the world that use the hematite process and are therefore already equipped with such autoclaves; all other plants use jarosite or goethite under atmospheric conditions and are therefore not equipped with such autoclaves [**56**].

* The following diagram shows the hematite process:

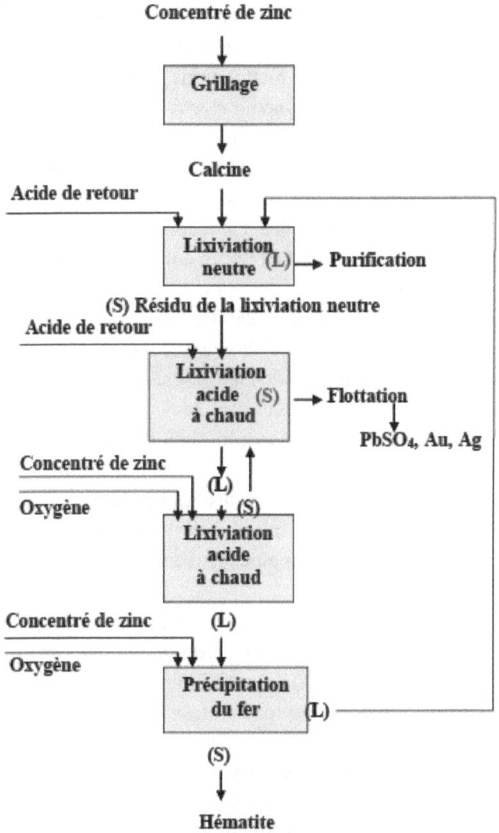

Figure 1.11: Hematite process **[57]**

1.4. Summary of scientific work on zinc extraction by hydrometallurgy

In zinc hydrometallurgy, sulphuric acid is the reagent used for leaching and decomposing roasted ores, concentrates or calcine. Iron is always present in raw materials as an impuriti , and must be removed from the leaching solution. Currently, approximately 75% of the world's zinc is produced by hydrometallurgy. Iron is generally precipitated in the form of jarosite, goethite or hematite.

The first two processes generate large quantities of residues, which are toxic due to their concentration of heavy metals (zinc and lead). The production of this type of waste in the European Union is estimated at 600,000 tonnes per year. To prevent leaching of the residues, they are stored in closed containers or in watertight tanks, which is a costly, environmentally friendly but unsatisfactory procedure **[58,59]**. In this context, several studies on iron precipitation have been demonstrated:
M.R.C. Ismael and J.M.R. Carvalho **[60]** proposed precipitating iron from aqueous zinc sulphate solutions in the form of pure hematite, which can be used as a pigment or raw material in the steel manufacturing industry. Pure hematite can be obtained if iron is removed from aqueous solutions using solvent extraction techniques such as organophosphoric acids, amino-phosphoric acids and amines. Although various extractants have been tested, the extraction solvent has not yet been successful in zinc hydrometallurgy [**31**].
Sárka Langová, Juraj Lesko, Dalibor Matysek **[60]** and M.K. Jha, V. Kumar, R.J. Singh **[61]** tested various leaching agents, namely hydrochloric acid, ammonia solution and sodium hydroxide, for zinc recovery. Another cathodic electrolytic process at room temperature was developed in India to recover zinc from industrial residues. Various parameters such as cyclic voltammetry and galvanostatic studies were used to illustrate the dissolution of zinc ferrites contained in the exhaust residue [**62**], unlike the Zinifex smelter in Hobart, which attributed this to the formation of a precipitate called Paragoethite (a process similar to the goethite process). the Paragoethite process is not well documented and its commercial use is limited to three sites worldwide [**63, 64**].
Other research **[65]** has developed a hydrometallurgical waste treatment process based on the destruction of the ferrite structure. To do this, ZnFe2Ü4 is treated with (FeCl3.6H2O). The reaction is carried out by exchanging $O(^{2-)}$ $/Cl(^{-)}$ ions'allowing zinc to be recovered as ZnCL and iron as Fe2Ü3 hematite. These products are separated by aqueous leaching. The final solid residue, which has been concentrated in iron, should be sent to the steel industry.
The implementation of the aforementioned processes for iron removal and zinc recovery from leaching residues in zinc hydrometallurgical plants has not yet been successful, although all of these processes have been tested on a pilot scale. The Jarosite, Goethite and Hematite processes remain the most widely used for zinc recovery.

Conclusion

Zinc is a transition element with atomic number Z=30 and belongs to group II B of the periodic table of elements. It ranks fourth after iron, aluminium and copper. It is relatively abundant in nature, where it is found in the form of sulphide called blende, but it also exists in other ores such as calamine. ZnS-based sulphide ores with a low zinc content of 2 to 12% are the raw material for zinc metallurgy. They undergo enrichment treatment by flotation before use, which is the first step in preparing the ore for zinc production.
Zinc differs significantly from other technological metals in its chemical, physical and mechanical properties , particularly its relatively low melting point (419°C) and its compact hexagonal crystal structure.
Due to its properties, zinc occupies an important place in the non-ferrous metals industry. It is used as a means of protecting steel by coating (galvanisation) and as laminar zinc in various branches of industry. Zinc is also used in corrosion protection, metallisation, cathodic protection, IT and battery manufacturing. Its chemical and physical characteristics allow it to be used in many applications while preserving the environment (natural environment). Pure zinc is not considered toxic and is 100% recyclable, making it a material that preserves the

environment and promotes sustainable development.
Extractive metallurgy can be divided into two parts: pyrometallurgy and hydrometallurgy. Pyrometallurgy was the first to be used, dating back to ancient times. Hydrometallurgy, on the other hand, only emerged at the end of the 19th century. Hydrometallurgy has developed significantly since the beginning of the century and has often overtaken pyrometallurgical processes for the production of many metals (Zn, Ni, Cu). Initially used for the treatment of ores or concentrates, it has been used for the treatment of metal waste for the past fifteen years.
The zinc contained in ores is never entirely recovered in ingot form, as a significant percentage remains dispersed in metallurgical operations in the form of residues from various products during the different stages of extraction. Zinc extraction processes are characterised by solid waste (metal sludge), gaseous emissions (SO2 emissions) and liquid waste (ion-laden water, acidic water, etc.), which have a negative impact on the environment and human health.
Several processes have been developed, such as Jarosite goethite and hematite, with the aim of increasing the dissolution rate in order to convert zinc into ionic form, on the one hand, and to recover zinc from solid waste, on the other. The double leaching process only achieves a dissolution rate of 90%, whereas with the Jarosite process, the rate can reach up to 96%.
In this chapter, we have presented the various properties of zinc, the stages of pyrometallurgical and hydrometallurgical processing, the processes for treating solution residues in pyrometallurgy and hydrometallurgy, and some proposed work on zinc extraction in hydrometallurgical processes. In the next chapter, we will present the industrial context of the double leaching process (ALZINC) and the environmental issues associated with solid leachate discharges.

CHAPTER 2

Industrial context of zinc extraction and its impact on the environment

Introduction

After presenting zinc in the first chapter, including its structure and properties and the various processes used to extract it from ores, this chapter will present the process of zinc production by hydrometallurgy and neutral and acid double leaching, which continues to be used at the ALZINC plant in Ghazaouet, discussing the various discharges resulting from this process and the issues of open-air storage or technical landfill (CET).

Figure 2.1: *Geographical location of the zinc production plant*

Since 1974, when the plant began operations, the Ghazaouet site was chosen for several reasons, including its proximity to the raw material (blende) from the El Abed mine (30 km from the plant), the availability of sea and rail transport, and the workforce.However, due to the inadequacy and subsequent closure of the El Abed mine, the plant now processes ore imported from Latin America (Peru) at a capacity of approximately 300 tonnes per day, producing 40,000 tonnes of zinc per year. The double leaching process only achieves a leaching rate of 90%. This results in a mass of leaching-insoluble material that constitutes a solid waste of metal sludge containing more than thirty chemical elements **[26].**

In this study, we are interested in the recovery of solid waste from leaching and purification in zinc hydrometallurgy, given its impact on the environment and the balance of ecosystems. Solid waste (leaching residues) contains harmful elements such as Pb, Zn, Cr, Cd and As. In 2001, special waste management was implemented through the construction of storage basins isolated by a polymer membrane to relocate the sludge stored on the cliff, estimated at over 400,000 tonnes, which posed a risk to the environment. This activity results in gaseous emissions of harmful particles such as SO2 [**66]** into the atmosphere and liquid emissions, which studies [**67]** have shown to be acidic process water that may contain heavy metals with significant signs of environmental degradation.

Several studies have been carried out in Algeria **[66-69]** with the aim of finding a solution to the management of solid, liquid and gaseous industrial waste.

In this work, we are interested in the recovery of solid waste from leachate in zinc hydrometallurgy. We will begin by describing the double leaching process, the various industrial waste streams from the zinc electrolysis plant and waste management, as well as their impact on the environment. We will then discuss the issues addressed in our study and conclude this chapter with a summary.

2.1 Zinc production using hydrometallurgical processes

The process involves several operations: roasting, leaching, purification and electrolysis.

The concentrates are transported and stored as follows: they arrive at the plant by sea and are taken to the storage area by lorries, which are weighed on a weighbridge to determine the quantity of concentrate entering the plant. The storage area consists of six uncovered compartments, each with a capacity of 3,000 tonnes, where the ore is stored according to the quality of the blende. From there, the ore is loaded onto a conveyor belt by crane and sent to six hoppers. Depending on the desired quality, the blend is directed to a conveyor belt that connects to another inclined conveyor that leads to the screen. Here, the fines and rejects are separated. The fines fall into the furnace feed hopper and, by means of a rotating plate on the projection belt, feed the furnace directly. The tailings go to a crusher where they are ground to obtain the desired particle size. At the bottom of the crusher, they are sent back to the screen by means of a bucket elevator. The same applies to the waste **[70]**.

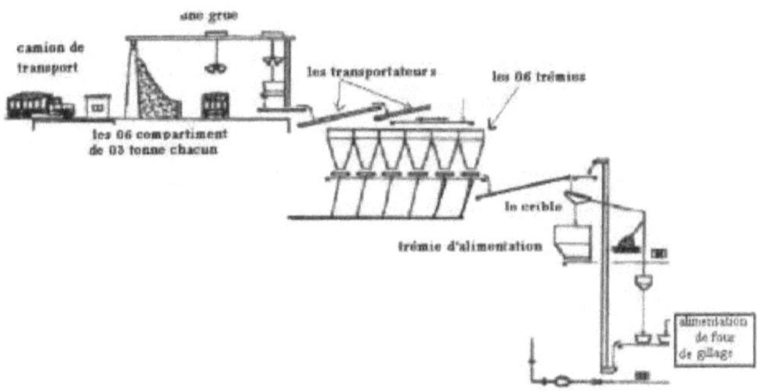

Figure 2.2: Transport and storage of raw material

2.1.1 Screening

The screening plant began production in 1975, with a processing capacity of 300 tonnes of ore per day. The first treatment to which the ore (ZnS) is subjected is oxidation to obtain calcined ZnO. This operation is carried out by heating to 900-950°C in the presence of air according to the following exothermic reaction:

$$ZnS + {}^{3}/_{2}\,O_2 \rightarrow ZnO + SO_2 + \text{Vapeur} \quad (2.1)$$

With a heat AH=-445 KJ/mole of ZnS between 800 and 1000°C

This form of zinc facilitates its wet extraction by acid and neutral leaching for the production of $ZnSO_4$, which will feed the electrolytic cells. Zinc oxide is soluble in sulphuric acid. The calcine produced is stored in two silos.

The roasting operation is carried out in a fluidised bed in a cylindrical furnace flared upwards' which has a 46 $^{m^2}$ grate and three coils to cool the fluidised bed. In the event of a shutdown, the furnace is heated by three diesel burners and six fuel lances.

The furnace is fed with blende from a feed hopper by means of a blende extractor, and the furnace is loaded by means of a rotating hearth to ensure a regular feed. The concentrate is distributed in the furnace by a loading machine. To achieve good desulphurisation of ZnS ore, the temperature inside the roasting furnace must be maintained at 900-950°C and an air flow rate for the oxidation of ZnS must be maintained to keep the fluidised bed in good fluidity conditions by means of a blower called the furnace blower **[70].**

The furnace is equipped with a heat recovery boiler and a gas purification system. In this process, part of the calcine escapes with the roasting gases at the furnace outlet. Heat is recovered by means of a boiler and the roasted ZnO is recovered from these gases by two cyclones and dry electrostatic precipitators. The recovered calcine, after treatment to adjust its particle size, is conveyed to two silos for subsequent use in leaching [**70].**

The roasted ZnO must meet strict requirements in order to be leached later. It must not contain more than 0.3% sulphur in the form of sulphide and 1.8% sulphur in the form of sulphate, and the formation of Zn ferrites ($ZnOFe_2O_3$) must be avoided by controlling the temperatures during roasting within the optimum values. The residence time of the blende particles entering the furnace is relatively long, which ensures that the calcined product has a low sulphur content in the form of sulphides and sulphates.

Secondary chemical reactions are likely to occur during the roasting process, mainly at lower temperatures. Sulphates form at 500–600°C and decompose at around 900°C.

$$ZnO + SO_2 + 1/2\ O_2 \longrightarrow ZnSO_4 \quad (2.2)$$

When the concentrate is rich in iron, which is often the case, ferrites, mixed oxides of zinc and iron, are formed according to the reaction:

$$ZnO + 2\ FeS + 7/2\ O_2 \longrightarrow ZnOFe_2O_3 + 2\ SO_2 \quad (2.3)$$

Other complex reactions may occur, such as iron sulphides or oxides of other elements such as copper, arsenic or lead, which can interfere with zinc extraction during leaching and originate from the raw material.

	$7FeS_2 + 6O_2$	$\rightarrow$	$Fe_7S_8 + 6O_2$	(2.4)
Fer	$Fe_7S_8 + O_2$	$\rightarrow$	$7FeS + SO_2$	(2.5)
	$3FeS + SO_2$	$\rightarrow$	$Fe_3O_4 + 3SO_2$	(2.6)
	$Cu_2S + 2O_2$	$\rightarrow$	$2CuO + SO_2$	(2.7)
Autres	$4As_2S_3 + 17O_2$	$\rightarrow$	$6As_2O_3 + 8SO_2$	(2.8)
	$2PbS + 3O_2$	$\rightarrow$	$2PbO + 2SO_2$	(2.9)

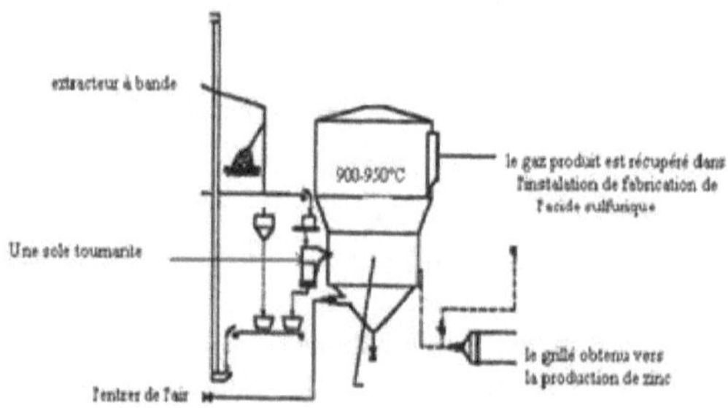

Figure 2.3: Roasting furnace

2.1.1.1 H_2SO_4 acid production

Sulphuric acid is a by-product of the combustion of blende, where the roasting operation releases a significant amount of SO_2 gas that requires adequate treatment before being released into the atmosphere. It will be used to manufacture sulphuric acid using another process in parallel with zinc extraction. The average production of sulphuric acid is 270 tonnes/day. In order for the treatment of roasting gases to be economical, it will be necessary to maintain the SO2 concentration at around 9-10%. After purification, the sulphurous gases from the roasting process are sent to contact systems to convert sulphur dioxide SO2 into concentrated sulphuric acid (98%) according to the following reactions:

$$SO_2 + {}^1/_2\,O_2 \rightarrow SO_3 \quad (2.10)$$

$$SO_3 + H_2O \rightarrow H_2SO_4 \quad (2.11)$$

The gas containing 8.5% SO_2 by volume enters the scrubbing tower to remove the remaining dust. It passes through a counterflow of sprayed water, where the gas is cooled to a temperature of 60°C and encounters a recycling liquor as it passes through a high-speed tray. Bubbles then form, allowing the water to remove most of the dust. The temperature drops to$^{38°C}$, and the liquor carries with it a quantity of sulphur dioxide and arrives at a degasser where the gas escapes into the atmosphere. The gas from the chimney mainly contains N2, O2, SO2 and a quantity of H2SO4 vapour.

In order to remove the vapours carried by SO_2, the latter passes through wet electrostatic precipitators.

The gas leaving the wet electrostatic precipitators passes through the drying tower where its moisture is removed by contact with a countercurrent of 96% sulphuric acid (with the addition of 96% H_2SO_4 to maintain the acid at 98%). The dry, clean SO2 passes through two heat exchangers arranged in series, where the temperature rises to$^{430°C}$at the outlet.

The sulphur dioxide enters the converter, which is a large-diameter reactor containing four beds of vanadium pentoxide (V_2O_5) catalyst.

The sulphur dioxide is oxidised to sulphuric anhydride.

$$SO_2 + 1/2O_2 \rightarrow SO_3 \quad (2.12)$$

As the reaction is exothermic, an air circulation coil is used to maintain the temperature inside the converter at 430 °C. After oxidation, the gas passes through cold heat exchangers, where SO3 reaches a temperature of 178 °C at the outlet and enters the absorption tower, where the sulphur dioxide is absorbed by contact with a counterflow of sulphuric acid at 96 °C to form sulphuric acid at 98 $^{°C}$.

$$SO_3 + H_2O \rightarrow H_2SO_4 \quad (2.13)$$

This reaction is accompanied by a large release of heat. The gases leaving the absorption tower are discharged to the chimney. The acid obtained is cooled and then sent to storage **[70]**.

2.1.1.2 Steam

This is a derivative produced during the roasting operation resulting from the combustion of the blende. It is used in the rest of the metallurgical process. In the event of a shutdown, steam is produced by an auxiliary boiler to ensure the smooth running of the plant.

2.1.2 Leaching

The main purpose of this stage is to dissolve as much zinc as possible in the form of zinc sulphate and prevent the dissolution of impurities in order to eliminate a large proportion of these impurities by oxidising ferrous iron into ferric iron, which acts as a carrier. This stage also allows solids to be removed by settling.

This process consists of neutral leaching carried out in the presence of excess roasted ore, followed by acid leaching to complete the dissolution of the zinc, which causes significant recirculation of impurities, particularly iron and copper. This operation is entirely continuous [71].

2.1.2.1 Roasted ore circuit

The zinc ore from the roasting furnace is stored in two 2,000-tonne silos before being sent to two pulping tanks. ZnO is soluble in dilute acid according to the following reaction:

$$ZnO + H_2SO_4 \rightarrow ZnSO_4 + H_2O + chaleur \quad (2.14)$$

a) Pulping tank

The roasted ore from the roasting furnace, after storage in two 2,000-tonne silos, is sent by screw conveyors to one of the two pulping tanks where the roasted ore (ZnO) is partially attacked by the solution from the mixing tank, where a mixture of solution and roasted ore is added simultaneously. The mixture is thoroughly stirred using an agitator to prevent solids from settling and to ensure greater homogeneity. The aim is to use the head acid and zinc oxide to form a pulp which will be dissolved by neutral leaching. A high concentration of copper in solution will attack the agitators, pumps, hydrocyclones, balls and even the inside of the mills. At the outlet of the pulping tank, the pH is 5.2 to 5.4 and we have:

- 20 to 25% of ZnO converted to $ZnSO_4$.
- 50 to 100 mg/l maximum copper.
- 200 to 250 g/l of solids **[71].**

b) Mixing tank solution

The mixing tank receives the head acid, the acid overflow, the filtrate from the rotary filters and, of course, a cell return additive.

The addition of cell return to the mixing tank is the first phase of the process. This allows the gradual dissolution of soluble zinc silicates and the precipitation of silica **[71].**

♦ **Head acid (preparation tank):**

The head acid consists of the purification return (neutral repulping O.F.) and the manganese solution.

♦ **Manganese tank:**

This is a mixture of the solution recovered after vacuum dredging of the electrolysis cells and purging of the anode washing tanks, to which manganese dioxide **(MnO_2)**, iron sulphate and cell return are added according to the reaction:

$$MnO_2 + 2H_2SO_4 + 2FeSO_4 \rightarrow Fe_2(SO_4)_3 + MnSO_4 + 2H_2O \quad (2.15)$$

♦ **♦♦ Grinder:**

The rejects from the screens and hydrocyclones pass through a crusher. This is a drum with an inner steel lining containing approximately 11 tonnes of iron pellets of different diameters (0.40, 0.50 and 0.60 mm). The presence of copper is harmful, as it not only attacks the iron balls but also risks infiltrating and attacking the drum lining.

In this case, it is necessary to regularly analyse the copper content and determine the filtrate of the solution leaving the grinder **[71].**

♦ **♦♦ Hydrocyclone**:

The solution leaving the Pulping tank is pumped to a hydrocyclone where initial separation (solid-liquid) takes place. Particles larger than **75 μ** are normally returned to the mill, while the fines are carried by the solution into the six neutral leaching tanks where they undergo a second attack.

The hydrocyclone regulates the smooth running of the entire leaching process. Solid-liquid separation depends on the flow rate of the solution and the capacity of the hydrocyclone.

At low flow rates, all the solution admitted into the hydrocyclone flows to the mill. At very low flow rates, there is a risk of blocking the lower outlet and consequently returning all the large particles to the neutral leaching tanks. In the latter case, leaching will be incomplete, resulting in a loss of zinc in the form of zinc oxide that has not been converted to **ZnSO4**, and there is a risk of blocking the tanks and decanters **[71].**

Figure 2.4: Leaching

2.1.2.2 Leaching parameters

The purpose of leaching is to transfer a chemical element from the solid phase (calcine) to the liquid phase (using an attacking solution). It involves physical transfer phenomena on the one hand and chemical transformation phenomena on the other **[71].**

- **Physical parameters**

The particle size of the solid, which has a direct influence on the exchange surface, plays an essential role in the kinetics of contact between the solid and liquid phases in solution. The degree of fragmentation (grinding) of the solid particles of the calcine is an important parameter.

Table 2.1: Particle size of ZnO (80 to 82% of the calcine has a particle size between 45 μ and 80 μ)

φ Grains	**< 0.045 mm**	**> 0.045 mm**	**> 0.080 mm**	**> 0.25 mm**	**> 0.65 mm**
ZnO %	**18**	**80 - 82**	**18**	**0.16**	**0**

- **Chemical parameters**

Zinc oxide is transformed during dissolution by the action of dilute sulphuric acid. The speed of this reaction generally depends on the concentration of this reagent, the grain size and the roasting temperature **[71].**

2.1.2.3 Neutral leaching

This is an initial continuous solution process in six 25 m³ tanks, where the pulp obtained is mixed with the return cell from zinc electrolysis, which has the following composition:

H_2SO_4 libre: 160g/l

Zn: 50g/l

Temperature: 55 to 60°C. The flow rate is regulated by two pH meters placed between the first and second tanks and between the second and third tanks to obtain a pH between 5.1 and 5.3.

The injection of air is essential to produce an upward movement of the solution and oxidise ferrous iron into ferric iron. The latter precipitates in the form of Fe(OH)3 hydroxide, carrying with it As, Ge, Sb, etc. In the fourth tank, a flocculant (magnafloc) must be added to agglomerate the solid particles and thus improve settling. The overflow from the neutral leaching is sent to three settling tanks, known as neutral settling tanks, each with a capacity of 400 m^3, which separate the solution to be purified from the sludge containing the ZnO that has not been attacked in the neutral leaching tanks.

The overflow is sent to purification and the underflow is sent directly to acid leaching **[71].**

- **Neutral settling tank**

Three decanters separate the solution to be purified from the sludge containing unattacked ZnO in the neutral leaching tanks.

The decanters are designed for a maximum iron content of 6% and a useful flow rate of 50 m^3/h.

This is because the amount of sludge is proportional to the iron content. High iron content leads to the formation of insoluble zinc ferrites, which not only causes zinc losses but also significant sludge deposits in the decanters; the neutral overflow becomes less clear and contains solid particles that cause filterability problems during purification. (Iron is reduced).

Each decanter is equipped with a scraper that directs the sludge towards the discharge opening.

The presence of a lot of sludge prevents the scraper from turning freely. This resistance is converted into torque. A pressure of 1 to 2 kg/cm^2 is tolerated on the torque measuring stop.

A scraper lifting system allows sludge that has settled on the arm to be removed for easier disposal.

The slightly greenish overflow containing only 0.2 g/l of solids is sent for purification.

The underflow is sent to acid leaching, where the sludge still rich in zinc is attacked in a more acidic environment **[71].**

- **Behaviour of impurities during neutral leaching:**

Roasted blende contains iron, copper, cadmium, cobalt, nickel, germanium, arsenic, lead, etc. as its main impurities. These behave as follows:

Iron: The iron contained in blende is practically insoluble under leaching conditions (iron in the form of ferrites or Fe_2O_3). However, ferrous iron must be added, which will be converted into ferric iron to maintain the Mn content of the electrolyte. This solution comes from electrolysis, where iron sulphate is oxidised by Mn02 according to the following reaction:

$$2FeSO_4 + MnO_2 + 2H_2SO_4 \longrightarrow Fe_2(SO_4)_3 + MnSO_4 + 2H_2O \quad (2.16)$$

By neutralising the attacking acid with roasted blende, the iron hydrolyses and precipitates almost completely, causing the other precipitates to be absorbed or form basic salts.

$$Fe_2(SO_4)_3 + 3\,ZnO + 3H_2O \longrightarrow 2Fe(OH)_3 + 3\,ZnSO_4 \quad (2.17)$$

The conversion of ferrous iron (**Fe^{2+}**) to ferric iron (**Fe^{3+}**) is activated by injecting air into the leaching tanks. The precipitation of **20 g** of iron produces approximately **1 g** of impurities.

Copper: Mainly in the form of CuO, copper precipitates partly as hydroxide, while the excess remains in solution. The precipitates are carried away by $Fe(OH)_3$ (40% remain soluble).

Cadmium and nickel: Cadmium and nickel pass into solution. They only precipitate as hydroxides at **pH values** above **8** and **7**, respectively.

Cobalt: Cobalt salts are highly unstable and are converted to $CoSO_4$. Cobalt passes into solution.

Germanium: Leaching only precipitates part of the germanium in the form of hydroxide.

Arsenic: As2O3 is converted into H3ÄsO4 or insoluble arsenates, most of which precipitate with iron. Removal is more complete than for germanium. Arsenic is used as a control for leaching and neutral settling operations.

Antimony: Antimony is effectively removed in the form of Sb(OH)s; insoluble antimoniates.

Manganese: Manganese is not removed and will remain in solution.

Selenium and tellurium: Highly harmful to electrolysis, they precipitate in the form of hydroxides and will be carried away by Fe(OH)3.

Silica: SiO_2 is insoluble.

$$SiO_2MeO + H_2SO_4 + 3H_2O \longrightarrow 3\ MeSO_4 + SiOO_4 + H_2O \quad (2.18)$$

They precipitate in the form of colloidal silica and are carried away by **Fe(OH)**3.

Lead: Trivalent and pentavalent salts are reduced to bivalent salts Pb^{++} , $PbSO_4$, which are insoluble and carried away with iron hydroxide.

Calcium: calcium is largely insoluble in the form of gypsum. However, its solubility is not negligible and solutions are supersaturated with calcium sulphate. This causes numerous undesirable crystallisations during manufacture **[71].**

2.1.2.4 Acid leaching

The underflow from the neutral decanters is treated a second time in four continuous tanks in , in a more acidic medium to recover more zinc (18%). This is based on the same principle as neutral leaching, with the addition of cell return, air injection and flocculant at the outlet of the fourth tank, except that the medium is more acidic (pH between 2.5 and 3).

The acid discharge solution is also sent to two acid decanters identical to those used in neutral leaching. The overflow joins the feed chute of the mixing tank and the underflow is sent directly to the rotary filter tanks and then sucked out by a vacuum pump and washed with hot water at 60°C to recover the $ZnSO_4$. The rotary filter filtrate passes through the cloth and returns to the head of the leaching tank, and the squeezed cake is removed from the cloth by blowing air and sent to the discharge [**71].**

- **Acid decanter**

These two decanters are identical to those used in neutral leaching. The acid overflow (earth-coloured) joins the feed chute of the pulping tank via the mixing tank. The acid underflow drawn from the bottom of the decanters is sent to the rotary filter tanks using diaphragm pumps (O.D.S).

The operating frequency of these pumps can be adjusted according to the pressure in the settling tanks. It also depends on the filter capacity of the rotary filter **[71].**

- **Filtration**

The acid leaching underflow contains approximately 500 ml of filtrate per litre, which must be recovered. This solution is pumped into the rotary filter tank and then sucked out by a vacuum pump. The rotary filter allows continuous filtration in short cycles.

The distributor is divided into several zones in its fixed part, each of which collects the different liquids (mother solutions, washing and rinsing solutions) from the suction pipes of the cells as they pass through during the rotation of the drum.

The sludge is washed with hot water (60°C) to recover the ZnSO4 (water-soluble zinc). The filter passes through the cloth and returns to the top of the leaching tank. The drained cake is removed from the cloth by blowing air at the scraper.

This operation is carried out continuously, in several very short stages:

1) Filtration corresponding to the submerged part of the drum.

2) Washing of the cake formed in the previous zone.

3) Spinning the cake.

4) Blowing and removal of the cake (unloading).

5) Neutral zone **[71].**

- **Behaviour of impurities during acid leaching**

Iron: Fe $(OH)_3$ will partially transform into $Fe_2(SO_4)_3$, which returns to the neutral leaching stage where it is most useful.

Copper: it will be partially returned to solution in the form of $CuSO_4$.

Lead: Lead will remain insoluble.

Arsenic: arsenates can be relatively soluble in acid and partially dissolve.

Calcium: $CaSO_2$ (gypsum) nevertheless remains in supersaturated solution **[71].**

- **Ore residue**

During the roasting process of zinc concentrates, iron combines with the remaining metals to form compounds known as "ferrites", which generally have the formula MeO Fe_2O_3, where Me represents any of the metals accompanying iron in the concentrates. These ferrites are insoluble in the aqueous sulphuric acid solution used in leaching. The higher the iron concentration in the ore, the greater the amount of metals that become insoluble during roasting. In the ore residues, zinc is present in four forms: - Insoluble ZnS.

- Zn (ZnO) soluble in H_2SO_4, insoluble in water. This loss is due to incomplete acid leaching or hydrocyclone failure (passage of large particles that are difficult to attack).
- Zn ($ZnSO_4$) soluble in water, loss due to insufficient washing with rotary filters. - ZnO.Fe_2O_3. Zinc ferrites insoluble under normal operating conditions. The cake containing iron and 18 to 20% zinc is sent to landfill **[71].**

Figure 2.5: Ore residue

2.1.3 Purification

The purification process receives the solution from leaching, as the overflow from the neutral decanters is too impure to undergo electrolysis. It undergoes a double stage of cementation and filtration. The clear solution is pumped to zinc electrolysis. The sludge is repulped to obtain a cadmium-containing solution which is stored **[72].**

Figure 2.6: Purification

2.1.3.1 Hot purification

This removes most of the impurities such as Ni, Co, Cu, Cd, Sb, Ge, Sn, Te, TL, etc., in order to obtain a good Faradaic efficiency during electrolysis.

The solution from the leaching process is first heated from 50-55°C to 75°C using two heat exchangers and superheated steam (200 to 260°C). At the outlet, the condensate is used to wash the acid leaching residues and to prepare the flocculants. The heated solution outlet is regulated by a flow meter and passes through a tank where CuSO4 and Sb reagents are dissolved.

In this tank, the copper content is maintained between 200 and 225 mg/l, while antimony is prepared in water before being added according to the Co content.

In the first treatment tank, zinc powder is added according to the cobalt content. The solution passes through four 25 m^3 tanks, where it is vigorously agitated before being sent to the filter presses. At the end of the hot purification process, the cobalt content must be less than 0.25 mg/l.

Cobalt is removed using zinc powder in the presence of antimony, which is highly insoluble under the required pH conditions. Its dissolution is facilitated by the presence of copper. These are then cemented together with the cobalt by the excess zinc **[72]**.

The hot purification solution is sent by two pumps to three automatic filter presses. The filter press cakes are collected and sent to the first neutral repulping treatment tank.

The clear filtrates are collected by a gutter and sent to cold purification **[72]**.

* The following diagram shows a simplified diagram of hot purification:

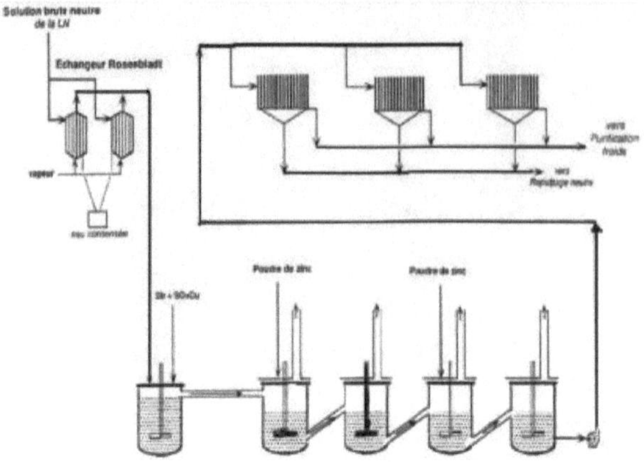

Figure 2.7: Hot purification

2.1.3.2 Cold purification

For a second cementation. It is useful to remove the last traces of thallium and cadmium. Monitoring the evolution of the latter indicates that this operation is working properly, as the cadmium content must be less than 0.2 mg/l at the end of cold purification.

The solution obtained from hot purification is sent to two tanks in series, each with a capacity of 25 m³. Fine or extra-fine zinc metal powder is added to the first treatment tank. At the outlet of the second tank, the solution is pumped into three 60 m³ manual filter presses[2] each, in order to separate the solids from the liquid by filtration. The clear filtrate is stored in six tanks to decant all the basic salts and obtain a purified solution, and the filter press cakes are sent to the first neutral repulping treatment tank **[72]**.

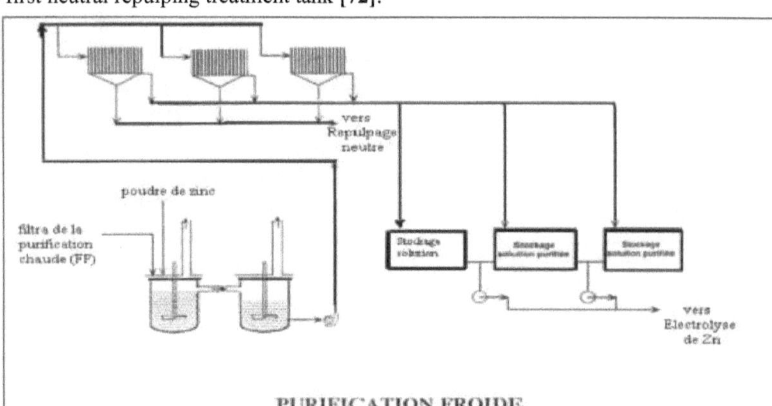

Figure 2.8: Cold purification

2.1.3.3 Repulping

The purpose of repulping is to recover, in a first stage, the excess zinc from hot and cold purification and, in a second stage, the cadmium contained in the same sludge **[72]**.

Figure 2.9: Repulping (according to Alzine)

• Neutral repulping

The purification filter cakes will be repulped in two 25 m^3 tanks and transported by screws feeding the first tank, where cells are added via a pH meter placed between the tanks. The pH is adjusted to between 5 and 5.1 so that only zinc is recovered and Cd, Co and Cu are not dissolved.

The solution, which passes from one tank to the other through a chimney that ensures better fluid circulation, is kept agitated before being sent by two pumps to a decanter. To decant the sludge properly, a Magna floc flocculant is added. The zinc-rich overflow (140 to 150 g/l) is recycled to the top of the leaching tank. The underflow passes through a rotary filter, the filtrate returns to neutral repulping and the Cd-rich sludge is reprocessed a second time in an acidic environment **[72].**

* The following diagram shows a simplified neutral repulping setup:

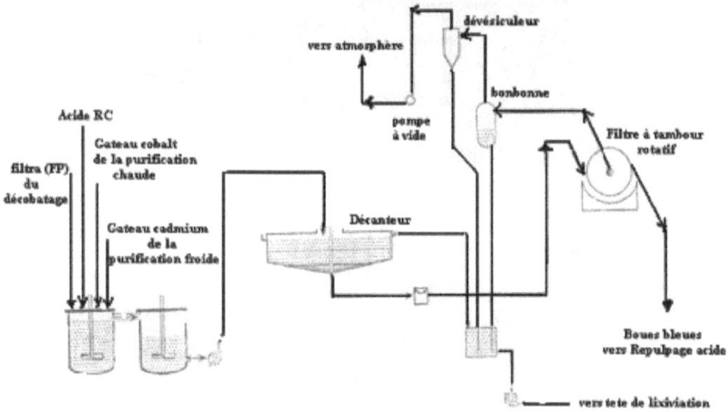

Figure 2.10: Neutral repulping

• Acid repulping

The blue sludge collected in the neutral repulping rotary filter is sent to three 25 m^3 treatment tanks, where cadmium is recovered in an acidic environment. The first tank receives the blue sludge from the rotary filter, the cell return (RC) from the zinc electrolysis and steam (to facilitate the dissolution of cadmium). The well-stirred and heated solution overflows from this tank through an internal circulation pipe connected to the second tank. A pH meter placed in this pipe controls the pH and regulates the addition of Cell Return.

In the second tank, the solution is kept at the same temperature by a second injection of steam. A second pH meter is placed at the outlet of this tank to maintain the pH. The last tank receives the solution from the second tank and the cloudy filtrates from the filter presses. This solution is recycled to the first tank and recirculates in a closed circuit until the end of the reaction. At the end of the operation, this solution is pumped into two filter presses. The filtrate (cadmium-containing solution) is collected in a 1.5 m^3 receiving tank and then pumped to the cadmium workshop. The copper sludge, known as black sludge, is manually discharged into a skip for storage.

N.B.: The cadmium workshop has been replaced by copper recovery since 2002. The workshop is currently shut down **[72]**.

* The following diagram shows a simplified acid repulping process:

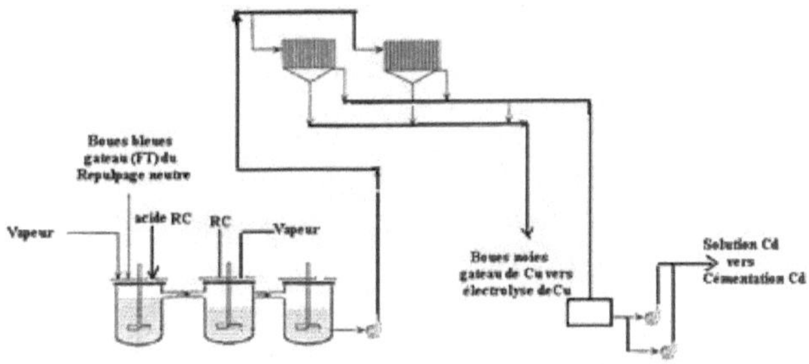

Figure 2.11: Acid repulping

2.1.4 Electrolysis

The main purpose of electrolysis is to convert zinc from sulphate form to metallic form by deposition, applying a potential difference between anodes and cathodes immersed in cells containing an acidic solution of ZnSÜ4. Zn^{++} ions are deposited on the aluminium cathode, while ÜH$^-$ions react at the silver-plated lead anode **[73]**. In dissociated bodies, the reactions are:

$$ZnSO_4 \rightleftharpoons SO_4^{-2} + Zn^{2+} \quad (2.19)$$
$$2(H_2O) \rightleftharpoons 2H^+ + 2OH^- \quad (2.20)$$
$$MnSO_4 \rightleftharpoons Mn^{2+} + SO4^{-2} \quad (2.21)$$

And the governing ion reactions are:

$$Zn^{+2} + 2e^- \longrightarrow Zn \text{ (cathodique)} \quad (2.22)$$
$$2(OH^-) \longrightarrow H_2O + 1/2\ O_2 + 2e^- \text{ (anodique)} \quad (2.23)$$
$$SO_4^{-2} + 2H^+ \longrightarrow H_2SO_4 \quad (2.24)$$

Overall reaction:

$$ZnSO_4 + H_2O \longrightarrow H_2SO_4 + Zn + 1/2\ O_2 \quad (2.25)$$
$$Mn^{2+} + 2\ H_2O \longrightarrow MnO_2 + 4H^+ + 2e^- \quad (2.26)$$

Zinc is deposited at the cathode and MnÜ2 at the anode.

And the operating conditions for the purified solution are:

- Temperature: 35-40°C.
- pH: 5.0 - 5.5.
- The current density is assumed to be: 375 A/m^3.
- The electrolysis time is 48 hours **[73]**.

Principle

The electrolysis workshop consists of two halls through which a current of 16,000 A passes. Each hall has nine rows of 24 cells with a useful capacity of 5.^{11} $^{m^3}$. They are made of reinforced concrete and each cell contains 40 cathodes and 41 anodes. The deposition time is 48 hours for each cell.

Figure 2.12: Electrolysis halls

The solution from the leaching/purification workshop is stored in six settling tanks to further promote purification, then transferred to the electrolysis workshop via an air cooler, which allows us to cool the purified solution. It then flows down a chute and comes into contact with the return flow.electrolysis workshop via an air cooler, which allows us to cool the purified solution. It then flows along a chute and comes into contact with the normal circuit return cell, which is generated by the six rows of electrolysis cells in a 150 m^3 and an additive in contact with the depleted cell return generated by the other three rows of electrolysis cells in another tank **[73].**

Figure 2.13: Cell return tanks

From these tanks, six pumps suck and discharge the two solutions into air coolers to cool the solution further to the optimum temperature for electrolysis. These air coolers consist of large chambers and fans that suck in air. At the outlet of these air coolers, the normal zinc-rich solution is divided into six ranges and the solution obtained from these ranges is called normal cell return. The depleted solution, which is low in zinc, returns to the depletion circuit, which consists of three ranges. The solution leaving this circuit is called the depleted cell return, and any surplus of this solution is returned to the head acid of the leaching/purification workshop. When the two solutions reach the electrolysis cells, the zinc ions are deposited at the cathode, where they are reduced to Zn, while the OH- ions are neutralised at the anode, with the release of oxygen.

oxygen. When the zinc deposit is considered sufficient, the electrodes are stripped, the zinc is collected at the cathodes, while at the anode, the manganese dioxide is stripped manually by scraping **[73]**.

Additives such as liquorice juice, gelatine and strontium carbonate are added to improve the electrolysis yield and for safety reasons:

- **Liquorice juice:** this is a foaming product that prevents excessive hydrogen release into the atmosphere.
- **Gelatin:** This prevents the deposit from coming loose during electrolysis by improving its adhesion. It also acts as a grain refiner, ensuring a smooth, even deposit.

- **Strontium carbonate** (SrCO3): This additive is used to remove lead that has passed into solution in order to protect the anodes.
- **Manganese**: Its purpose is to protect the anodes **[73]**.

> **Faradaic efficiency**

Faradic efficiency is the ratio between the deposited mass and the theoretical mass, which depends mainly on the hydrogen evolution that takes place at the cathode. This hydrogen evolution is due in particular to the zinc content of the electrolysis, the acidity and the presence of impurities **[73]**.

2.1.5 Remelting

Remelting is the final stage in the zinc metal production chain and consists of the following workshops:

- Zinc remelting workshop.
- Zinc powder workshop.

The zinc from the electrolysis department is transferred to the storage area on loading boards after being weighed. It is dried for approximately one hour to remove moisture **[73]**.

2.1.5.1 Zinc remelting

The zinc is loaded into two parallel charging columns that are in direct contact with two induction furnaces operating at a temperature of 450 to 500°C and with a capacity of 40 tonnes each. Once the zinc has melted, The bonding operation is carried out using a compressed air zinc pump. The ingot chain, consisting of 160 ingots, transports the zinc into moulds, which are then cooled by a cooling conveyor. An automatic stacking machine receives the zinc, which is then stacked in piles of 40 ingots weighing 25 kg each. It is then weighed using an electric scale and stored in batches for strapping. It should also be noted that there is a set of two one-tonne moulds and two two-tonne moulds that operate according to the order [**73**].

Figure 2.14: Remelting workshop

2.1.5.2 Zinc powder workshop

Depending on the process requirements (purification), the induction melting furnace, which is connected to a holding furnace and three electric resistance crucibles, is filled. The molten zinc flows by gravity to be sprayed into a chamber. The pulverised zinc is sucked into compartments equipped with special filters where a beating system is used to recover the zinc powder dust (extra fine) [**73**].

2.1.5.3 Quality of the zinc produced

The zinc metal product is of high purity, which has enabled it to be registered on the London Metal Exchange under the "**SNS SHG**" label. Two-thirds of the zinc placed on the world market must be at least **99.995%** zinc. Quality has become the benchmark for competition and the main catalyst for progress and development in the various areas of the company's activity. To this end, a laboratory carries out analyses on all by-products during the process, from raw materials to finished products, in order to guarantee quality that complies with standards.

In order to achieve its quality objectives, the company has equipped the laboratory with high-tech in- equipment and has made staff aware of the importance of quality for the company's long-term future **[44]**.

2.2 Types of waste generated during zinc production

Zinc production is accompanied by a series of emissions that can be classified according to their nature as follows:

- Gaseous waste
- Liquid emissions
- Solid waste

2.2.1 Atmospheric emissions

At the main factory chimney, despite being treated in the acid contact workshop for sulphuric acid production, the gases generated by the oxidation of zinc sulphides retain certain concentrations of nitrogen oxides and H_2SO_4. The Ghazaouet sulphuric acid plant operates by simple catalysis and cannot exceed a conversion rate of 97.5%, which, with an inlet gas containing 6% SO_2 by volume, results in a SO_2 content of 0.15% in the emissions **[66].**

In 1988, work carried out at Alzinc concluded that during normal production, SO_2 concentrations released into the atmosphere were 0.15%, which is acceptable and below international standards; however, the problem arises during shutdowns and restarts of the manufacturing process, when pollution peaks cannot be avoided. Before 1990, shutdowns were mainly due to water cuts, with 77 cases reported over a 13-month period. The water problem was subsequently resolved by the construction of a seawater desalination plant, which prevented frequent shutdowns. However, it should be noted that an average of two to three interventions are carried out each year, plus a general shutdown that is scheduled each year in August (corresponding to the workers' holiday period). The various shutdowns (excluding the annual shutdown) are caused by either:

- A leak in the pipework.
- Defective parts need to be replaced.
- Ore sticking to the furnaces.
- Other **[66].**

Solutions have been considered with the aim of minimising or even neutralising gaseous effluents through the use of coram solutions for gas treatment and the refurbishment of roasting equipment, as well as the continuous supply of water from the newly acquired desalination plant.

2.2.1.1 Gas treatment

One solution for neutralising gas emissions is a treatment that reduces SO_2 emissions into the atmosphere to no more than 400 ppm (0.04%) for an intake of 1500 ppm *(0.15%)* **[66]**. This is achieved by washing the gases from the acid workshop with a lime milk solution.

Another solution is to refurbish the roasting equipment (absorption tower and drying tower). The replacement of the gas absorption and drying towers, which were in a very advanced state of corrosion, had become necessary to ensure the proper functioning of the acid workshop and the waste gas treatment plant. Gradual commissioning took place on **1 April 2003** after the refurbishment of the roasting and acid workshops **[66].**

2.2.1.2 Continuous water supply

Faced with difficulties in supplying the complex with raw water and frequent shutdowns and restart due to water unavailability, which were a source of pollution due to the destabilisation of all technological processes, the company built a seawater desalination unit with a capacity of 2,500 m^3 per day **[66]** using a multi-effect distillation and vacuum vapour compression process.

2.2.2. Liquid discharges

The factory's liquid waste contains a considerable amount of complex chemicals, the impact of which on the marine environment is difficult to predict, especially as it is exacerbated by the cumulative effects they can cause. In addition to liquid waste, the plant uses pre-chlorinated water to cool the sulphuric acid production facilities and parts of the thermal power plant.

The volume of water discharged reaches approximately 300 m^3/h, or 72,000 m^3/day, as the plant operates 24 hours a day. These liquid effluents come from:

- The raw water demineralisation plant.
- The gas purification tower.
- Acid purging (absorption and drying tower)
- Overflow from cleaning the anodes and cathodes in the electrolysis workshop.
- Suspicious water collected in the plant, such as contaminated rainwater, etc. **[66].**

The solutions implemented for liquid waste consist of the commissioning in 1987 of an acid liquid waste neutralisation plant. The purpose of the neutralisation plant is to treat the factory's acid effluents in order to render the metals insoluble and bring the effluents to a **pH** level that allows them to be discharged into the sea.

The effluents consist of:

- Acid effluents from demineralisation.
- Liquid effluents from the treatment plant (screening).
- Other

The effluents stored at the head of the neutralisation unit are treated by adding lime milk or:

- Underflow will be filtered on a drum filter. The cakes (hydrated gypsum is precipitated) will be transferred for disposal
- Overflow and filter cake can be discharged to the sewer. Part of the clear solution is recycled to the head for lime milk preparation **[66].**

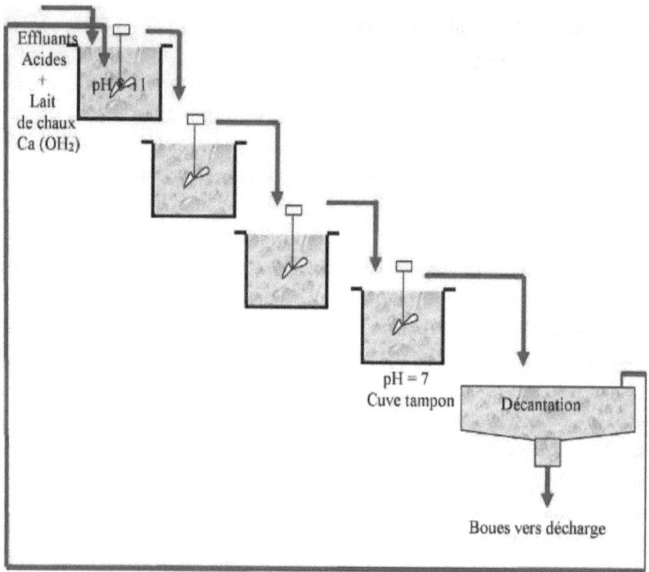

Figure 2. 15: Liquid effluent neutralisation station

2.2.3 Solid waste

The main industrial wastes produced by the Ghazaouet zinc electrolysis plant are:

- Copper sludge
- Leaching residues

2.2.3.1 Copper sludge

They are obtained in the purification workshop and were estimated in 1992 to be approximately 6,500 tonnes, with a production rate of 37 tonnes/month. They are stored in a landfill within the factory and their chemical composition is as follows:

Table 2.2: Chemical composition of copper sludge [66]

Chemical compounds	Content %	Chemical compounds	Content %
Cu	29.69	Al	0.
Zn	12.9	Ni	0.12
Pb	9.96	As	0.02
S(SO4)	8.5	Ge	0.035
Cd	3.75	Ag	0.0075
Co	0.32	Tl	0.0075
Sb	0.2	Cl	0.054
Mn	0.15	H	17.7
Fe	0.1		

Figure 2.16: Storage of copper sludge in the open air

2.2.3.2 Leaching residues

These residues are in the form of a sludge with 40% moisture content, chocolate brown in colour, very viscous and heavy. When discharged, they flow fairly easily, but like a paste rather than a liquid. After exposure to air, they dry on the surface and become covered with a crust and a whitish film due to the deposition of zinc sulphate and gypsum **[66].**

Table 2.3: Chemical composition of leaching residue [74]

Content in %	**01/2008**	**02**
Zn	20.68	20.63
Zn(H_2SO_4)	9.55	9.9
Zn(H_2O)	6	6
Cd	0.16	0.14
Fe	26.35	25.38
P	4	4
Cu	0.82	0.85
Mn	0.44	0.63
Cl	0.014	0.014
F	0.06	0.05
Mg	0.77	0.73
Co	0.0057	0.007
As	0.05	0.05
Ni	0.0036	0.0038
Ca	2.86	1.33
Ge	0.013	0.013
Cr	0.008	0.0069
Tl	0.001	0.001

The elements present in ores that are insoluble in sulphuric acid are: silica, lead sulphates, barium sulphates, lime sulphates and silver sulphates. Zinc combined with iron in the form of zinc ferrites, which form during roasting at 900°C, are insoluble unless very high acid concentrations and temperatures are used: they are therefore stable and do not risk decomposing or dissolving under normal atmospheric conditions. Hydrolysable elements are removed at pH 4 and pH 5 by the precipitation of iron hydrolysis. These include aluminium, antimony, ferric iron and tin. Metals dissolved during leaching are not sufficiently removed by washing: soluble zinc, copper, cadmium, cobalt and nickel.

> **Formation of zinc ferrites**

Wet extraction of zinc by leaching is a technological process that involves passing zinc through a solution (dissolution), which can be done in a neutral or acidic environment. This transformation is not complete and results in solid residues containing more than 21% zinc and other elements. These residues may contain zinc in forms that are insoluble in the aqueous sulphuric acid solution used in hydrometallurgical operations. Their presence has long been an obstacle to these operations and they form when the ore concentrate is rich in iron, which is often the case. Ferrites, mixed oxides of zinc and iron, are formed according to the reaction:

$$ZnO + 2\,FeS + 7/2\,O_2 \longrightarrow ZnOFe_2O_3 + 2\,SO_2 \quad (2.27)$$

The higher the iron concentration in the ore, the greater the amount of metals that become insoluble during roasting. During the roasting process of zinc concentrates, iron combines with the other metals present to form compounds known as "ferrites", which generally have the formula MeOFe2O3, where Me represents any of the metals accompanying iron in the concentrates.

In hydrometallurgical residues, zinc can be found in four forms.

- Insoluble ZnS.
- Zn (ZnO) soluble in **H2SO4**, insoluble in water.
- Zn (ZnSO4) soluble in water.
- ZnO.Fe2O3. Zinc ferrites insoluble under normal conditions.

The accumulation of solid waste from storage is due to the formation of zinc ferrite complexes and metal sludge. Since 1974, when the zinc refinery (Algeria) began production, insoluble residues from leaching and purification of the electrolytic solution have been stored in the open air near the Mediterranean coast. Since 2001, special waste has been stored in basins isolated by polymer membranes.

Analyses carried out on the stored sludge have shown a diverse composition of chemical elements in varying proportions, the most important of which are iron (30%) and zinc (23%), existing in several forms, particularly zinc ferrites. Copper and cadmium, whose contents are less than 1%, have been extracted from the sludge for possible commercialisation. Heavy metals, such as lead (5%) contained in the stored sludge, are a source of pollution for the natural environment, particularly the coastline and groundwater.

2.3 Solutions for zinc extraction waste

The solutions provided for solid waste are isolated landfill storage (CET) and rewashing of residues.

In 2001, special waste management was implemented with the construction of storage basins for solid waste using isolated polymer membrane burial to relocate sludge stored on the cliff, estimated at over 450,000 tonnes, which posed a risk to the environment. This solution is a temporary, costly and environmentally unsatisfactory procedure because the pollution of the subsoil and groundwater by metal- s is only a matter of time, as these ponds are guaranteed for a limited number of years **[34,40]**.

A second solution consists of recovering 1 to 2% of the zinc contained in the leaching residues by rewashing the sludge with process water at a temperature of

60°C while preparing two tanks for mixing and two decanters. However, this operation remains insufficient as the sludge contains more than 20% zinc. This operation is currently being started (2009).

2.4 Impact of discharges on the environment

To conclude this chapter, we present the impact of gaseous, liquid and solid discharges on the environment.

2.4.1 Gaseous emissions

Since the plant began operating in 1974, the population of the city of Ghazaouet and doctors in both the public and private sectors have complained of an increase in respiratory, osteoarticular, dermatological and ophthalmological diseases. A toxicological study of the pollutants generated in Ghazaouet confirms the potential for the development of these and other diseases, in addition to the carcinogenic effects that may be caused by cadmium, zinc, chromium, etc., present in the smoke dust during plant shutdowns and restarts.

Chemical reactions in the atmosphere produce sulphuric and nitric acids from sulphur dioxide and nitrogen oxides. These acids reach the ground in the form of dry particles in rain and other forms of precipitation. The various reactions that follow in the soil alter the chemical composition of the acidified rainwater before it reaches reservoirs and watercourses: for example, these acids can dissolve clay and release aluminium ions, damaging vegetation and disrupting aquatic life.

In fact, pollution from the factory in the form of nitrogen and sulphur oxides has damaged various herbaceous strata in the Ghazaouet region, particularly shrub vegetation, which has been severely affected (e.g. the decline of trees at the C.E.M Commandant Ziani located opposite the factory) **[66]**.

2.4.2 Liquid discharges

The marine pollution generated by the plant originates from:

- Liquid discharges, especially when the neutralisation station is not operating, - The use of seawater as a cooling agent for the facilities,
- Liquid chlorine injected into the seawater intake for the plant's cooling system,
- Accidental spills of H2SÜ4,
- Acid discharges, however low, are a hazard. Experiments have shown that plankton is killed at pH = 6.5. However, the normal pH of seawater is 8.2 to 8.3, which is slightly alkaline, but it is a highly buffered environment that easily compensates for pH variations. Nevertheless, acid discharges can alter the carbon dioxide system, which is primarily responsible for the buffering capacity of seawater. In addition, underwater dives carried out around the plant have revealed an absence of marine life.

The use of seawater as a cooling agent in installations produces heat-laden water, which is returned to the sea, causing thermal pollution that has two types of consequences for coastal marine life:

A- Some result from the conditions of transit through the facility and only affect planktonic organisms contained in the water that is pumped.

B- the other consequences correspond to the alteration of the receiving environment resulting from its heating, which can kill 8% of plankton if the effluent temperature exceeds 40°C.

At sea, fish that have adapted to pollution are generally contaminated (work carried out by ISMAL) and remain subject to the accumulation of metals and heavy metals. Heavy metals, particularly toxic ones, accumulate in fish and contaminate humans with Cd, Zn, Pb and mercury in particular. The study of the toxicology of pollutants highlights and raises serious concerns about the development of various diseases among the inhabitants of the region, which must be the subject of specific studies and monitoring by competent medical institutions [**66**].

2.4.3 Solid waste

The pollution generated by the factory originates from solid waste, particularly sludge that has been stored in an illegal landfill located upstream from the factory on cliffs overlooking the sea. Runoff has led to the accumulation of metals and heavy metals, which contaminate fish and subsequently humans (same process as liquid discharges). The port of Ghazaouet has been affected by pollution from the factory. A study carried out by the Tlemcen Wilaya Environmental Inspection Office in 1991, with a view to the planned dumping of dredged waste from the port, quantified the degree of contamination of the port by the factory. The impact of the factory's discharges, combined with the maritime transport of minerals, generated an estimated 460,000 tonnes of dry matter. The consequences of this pollution of the port have necessitated the construction of a coastal dam at an estimated cost of more than 20 billion centimes and a delay in dredging work of nearly three years (until the coastal dam is completed), resulting in a loss of revenue for the port company of several million dollars annually. As for the port's grain business, only small grain ships can enter the port. The loss of revenue is estimated at $4 million per year [**66**].

2.5 Research questions and scientific approach

Given the inadequate management of special waste, the study within the framework of this project consists of recovering the metallic elements contained in solid leaching waste. It should be noted that the project only deals with leaching sludge; the treatment of gaseous or liquid effluents will be the subject of another study.

Effluents generated by zinc production facilities are generally contaminated with compounds that have a very high **environmental impact**, such as mercury, cadmium, lead, etc. Even at very low concentrations, these compounds are harmful to the environment.

While solutions have been found for gas emissions through the treatment of gases such as SO_2 and the rehabilitation of roasting equipment that was severely corroded, and solutions for neutralising liquid effluents through the commissioning of a neutralisation plant, solutions for solid emissions are insufficient, as open-air storage or isolated burial cannot provide effective solutions for the management of special waste, where zinc losses are considerable, in the order of 25% in addition to the harmful effect on the environment.

The aim of our work is to provide a technical solution through the successive recovery of all the elements contained in leachate sludge. We will begin initial work on the recovery of zinc using double leaching extraction processes followed by the Goethite process. These residues, which are rich in zinc and other elements, have the major disadvantage of being difficult to extract using the double leaching process and also pose a threat to the environment if stored in the open air. The aim is therefore to find a use for these residues by transforming them in a simple and economical way into an iron oxide-based pigment, which is a product of great industrial and commercial value with applications in many fields such as the paint industry. It should be noted that this work follows on from an engineering thesis [**44**] in which the student, using the double leaching process followed by the Jarosite process, was able to recover up to 14% of zinc from the overall content in the sludge of 23%, i.e. a recovery rate of 60.89%.

Conclusion

The main environmental problems associated with this industrial context of zinc production by hydrometallurgy are gaseous emissions from roasting furnaces, liquid effluents and

leachate solids.
Purges from flue gas cleaning systems are generally the most contaminated effluents generated by a smelter, although runoff and contact cooling water used to cool sulphuric acid production facilities and parts of the thermal power plant may also be contaminated. In the hydrometallurgical process, liquid effluents come from the raw water demineralisation plant, the acid purge roasting gas purification tower, etc., where atmospheric emissions are transferred to liquid discharges into aquatic environments.
Solid waste from zinc refining not only represents a loss in overall zinc yield estimated at over 23%, due to a 90% dissolution rate and the formation of zinc ferrites, which form complex compounds that prevent the zinc from being released into solution. These wastes form an accumulation of metal sludge stored in the open air or in technical landfills, causing damage to the environment and
ecosystem balance. In these residues, zinc can be found in four forms: insoluble ZnS, Zn (ZnO) soluble in H_2SO_4, insoluble in water, Zn ($ZnSO_4$) soluble in water, and $ZnO.Fe_2O_3$ zinc ferrites insoluble under normal conditions. Zinc ferrites are often difficult to extract in solution using the double leaching process. It will be necessary to supplement wet extraction with another iron complexation process such as the Jarosite, Goethite or hematite processes, which involve forming an easily filterable iron complex. In the Goethite precipitate, which is the subject of our study, the solid obtained can be easily and economically transformed into an iron oxide-based pigment, which is a product of great industrial and commercial value with applications in many fields such as the paint industry.
For gas emissions, various solutions can be implemented, such as treating gas emissions using purification systems, absorption towers, gas drying towers and refurbishing screening workshops.
For liquid discharges, in order to minimise the environmental impact of zinc refining, particularly the volume of effluent discharged, the neutralisation plant treats acidic effluent. The performance and reliability of the plant are crucial, given the potential environmental and financial impact of any non-compliance.
The recovery of 1 to 2% of the zinc contained in the leaching residues by rewashing the sludge with process water remains insufficient, as the sludge contains more than 20% zinc and other elements, in addition to the extraction by double leaching.
After presenting the hydrometallurgical zinc production process, the various plant effluen , their environmental impacts and technical solutions, Given the zinc discharge rate and the inadequacy of the processes used, in the next chapter we will develop a new method for extracting zinc from residues, which consists of using double acid and neutral leaching, followed by precipitation of iron in the form of goethite. This will release the zinc into solution and form a solid complex rich in iron, which will be recovered as a colouring agent. In the next chapter, we will present the different characterisation methods used, the methodology of the work and the results obtained during our work.

CHAPTER 3

Materials and methods

Introduction

The double acid and neutral leaching process (zinc dissolution) in zinc refineries is not complete, with only a 90% extraction rate achieved, resulting in solid metal waste in the form of ferrites that are insoluble in acid. The insoluble residues from leaching and purification of the electrolytic solution are sent to landfill. In this study, we present a technical solution for recycling zinc from a pulp prepared from leaching residues using the "Goethite" process. The aim is to find a use for the latter by transforming it in a simple and economical way into an iron-based pigment, which is a product of great industrial and commercial value with applications in many fields such as the paint industry.

The methodology we have used consists of taking samples of solid residues from the leaching plant's waste stream and carrying out laboratory tests using a double extraction method:

- First acid extraction using a solution of sulphuric acid from the return from the electrolysis cells (RC) and concentrated sulphuric acid as the attacking solution. We will carry out tests at different temperatures and pH values, modifying the reaction kinetics by varying the time.
- Extraction by double leaching followed by the Goethite process stages using a prepared sulphuric acid solution (return from electrolysis cells RC + H2SO4) as the attack solution, taking into account the physical and chemical conditions of the reactions.
- Recovery of Goethite precipitate by transforming it in a simple and economical way into an iron oxide-based pigment.

3.1 Materials and methods

The samples intended for analysis and treatment (leaching residue "rotary filter sludge", cell return) were carefully collected from different sites of the leaching plant and are representative of the mass of solid residues after double leaching. The samples were microanalysed at the ALZINC central laboratory using basic analytical techniques: volumetric, gravimetric and atomic absorption to determine the chemical characteristics of the elements extracted in each phase of the goethite and hematite process. Finally, analysis techniques specifically designed to study the quality of the final products were used, including Fourier transform infrared spectrometry (FTIR), inductively coupled plasma atomic emission spectrometry and X-ray diffraction.

3.1.1 Sampling

The sample taken undergoes a series of operations to obtain a representative sample on which all laboratory tests will be carried out. The sample taken is sieved on a vibrating screen with a particle size calibration of 4 mm after grinding to make the material as fine as possible for laboratory testing. The quartering method was used for sampling.

3.1.2 Moisture content

An initial test was carried out on the sample taken, which consisted of determining the moisture content of the residue using the direct measurement method by weight loss. The test samples were weighed after drying in an oven for 24 hours at a temperature of 105°C. The difference in weight before and after evaporation was the water content of the residue, expressed as a percentage. A desiccator was used to store the samples after the study and before weighing to prevent them from absorbing moisture from the air due to heat. Despite these precautions, this method did not produce a constant weight after evaporation. The results showed that the moisture content is approximately 40.9% for the tests carried out.

3.2 Analysis methods

Laboratory analyses were performed by volumetric, gravimetric and atomic absorption methods. Fourier transform infrared spectrometry (FTIR), inductively coupled plasma atomic emission spectrometry and X-ray diffraction were used to identify the quality of the Goethite and Hematite iron precipitates. We will provide a brief description of these methods and the equipment used.

3.2.1 Gravimetric methods

Principle

This quantitative analysis method aims to obtain a quantitative (i.e. total) separation of a cation or anion (in aqueous solution) by selective precipitation of an insoluble species. The method consists of transforming a soluble species into an insoluble compound in a specific medium. The precipitate formed is weighed, which allows the soluble species in the initial medium to be quantified.

The chemical species will have the following properties:

- it will not be soluble in the reaction medium.
- it must not absorb other ions or an unknown quantity of solvent.
- To facilitate filtration, given its isolation, it must crystallise well. **[75,76]** Gravimetric methods are classified into two main groups:
- Precipitation methods: the analyte is converted into a highly insoluble precipitate, which is then filtered, washed to remove impurities and transformed into a product of known composition by appropriate heat treatment. This product is then weighed.
- Volatilisation methods: the analyte or its decomposition products are volatilised at an appropriate temperature. The volatile product is collected and weighed, or the mass of the sample is weighed [76]. For gravimetric analyses'an analytical balance with an accuracy of IO"($^{4)}$was used'This device was carefully calibrated and cleaned before and after each measurement.

3.2.2 Volumetric methods

Principle:

This method of analysis is based on the exact measurement of the volume of the reagent solution, the concentration and the preparation of a titrated solution. Volumetric analysis is of great practical interest and has the advantage of being quick to perform.

The concentration of an aqueous solution containing an acid or a base is determined by volumetric titration. To achieve this, an acid-base reaction is carried out, which is generally chosen to be rapid and as complete as possible. At equilibrium, the acid and base reactants are mixed in stoichiometric proportions determined by the reaction equation used.

If Ca and Cb are the concentrations of the reacting acid and base solutions, respectively, and Va and Vb are the volumes of monoacid and base mixed at equivalence, we have the following relationship:

$$C_aV_a = C_bV_b \qquad (3.1)$$

This relationship expresses acid-base equivalence. To experimentally determine the end of the titration, we use:

- Either pH = f (Reagents) graphs using a pH meter.
- Or the colour change of carefully and precisely chosen indicators whose pH variation at equivalence is abrupt.

Titration by the complexometric method

The complexometric method using EDTA is used to determine the zinc content in all raw or roasted blends and, in general, in all materials containing high quantities of zinc. The complexing agent used is ethylenediaminetetraacetic acid (EDTA), with the following formula:

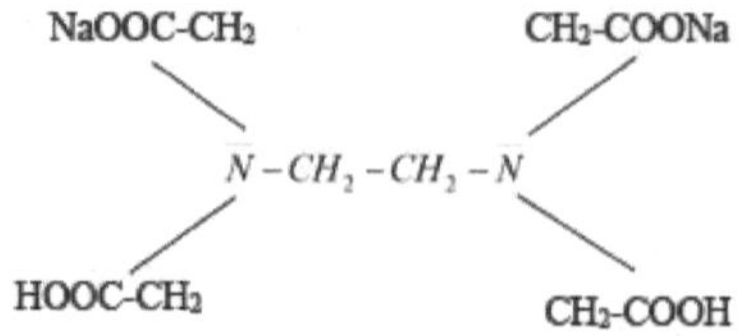

EDTA forms quantitative complexes with many metals, particularly Zn^{2+} according to the following complexation reaction:

$$M^{2+} + H_2Y^{2-} \rightleftharpoons MY^{2-} + 2H^+ \qquad (3.2)$$

M = Zn

Complex (Y'^{2}- E.D.T.A)

The release of H^+ protons shows that the position of this equilibrium depends on the *pH*. This must be fixed by adding a buffer solution [77].

3.2.3 Atomic absorption methods

Basic principles

Atomic absorption spectroscopy is based on the principle that a population of atoms in the Eo state can absorb photons of energy hv and that an estimate of the number of photons absorbed can be linked to the concentration of the element in the solution to be analysed. A population of atoms is generated in an atomiser. This population is illuminated by light radiation with a wavelength vo = 1/Ei · c and intensity Io. As this radiation passes through the atomic cloud, the atoms at the ground state Eo can absorb light such that, at the exit of the cloud, the light intensity is equal to I. The wavelength has not changed. The atoms that have passed into the excited state Ei very quickly (IO'^{5} to IO''^{9} s) to return to the ground state by emitting a photon with the same energy as the one that was absorbed and, therefore, at the same wavelength vo. This emission occurs in all directions and in particular at 90° to the incident radiation. It has an intensity F.

The absorbance A is defined as follows:

$$A = \lg (Io/I)$$

And it has been shown that absorbance obeys a law of proportionality similar to Beer-Lambert's law; this law allows two quantities to be defined:

- Transmittance *T*: expressed as a percentage:

$T = I/I_0 = e^{-\varepsilon l.c}$

$A = \log (I/I_0) = \varepsilon . l . c$

- Absorbance A:

Io Incident beam intensity.

I Intensity of radiation after passing through the heat source.

ε Molar extinction coefficient.

l Length of the optical path.

C Concentration of the solution to be analysed [78].

Equipment

An ***AVRORA*** air/acetylene flame atomic absorption spectrophotometer, connected to a microcomputer, was used in our analyses. It comprises the following components:

- A light source (primary source) that produces radiation characteristic of the element to be measured at the wavelength vo;
- A system for modulating the radiation from the source;
- An atomiser consisting of a nebulisation device, a burner and a flame, whose role is to produce a cloud of atoms in their ground state;
- A monochromator, which is used to eliminate all radiation other than that at the wavelength vo;
- A detector coupled to an electronic system for recording and processing signals.
- A light beam splitter **[78];**

Principle

An alternating current system is used as the energy source. The modulator consists of a rotating sector mirror that alternately passes the beam from the lamp through the flame (sample beam) and outside the flame (reference beam). The sample and reference beams are recombined by a semi-silvered mirror and pass together through the rest of the optics. The electronics are designed to measure the ratio of the beams, with the reading being the ratio of the energy of the sample beam in the numerator to the energy of the reference beam in the denominator. The two beams are recombined at the entrance to the monochromator. The detector receives these two beams alternately. The intensity of the reference beam is taken as Io [**79].**

Calibration

Atomic absorption spectrometry is a comparative analytical method that requires calibration, and the quality of the results depends on the representativeness of the standards in relation to the samples.

Calibration is obtained by measuring the absorbance of synthetic solutions with progressive concentrations of the analyte. The concentration of the unknown solution is then directly deduced by plotting its absorbance value on a previously established calibration curve. The standards are first prepared from a stock solution of 1 g per litre for each element and then sampled according to a standard operating procedure (for atomic absorption analysis) [**78].**

Preparation of test samples: for transfer to the apparatus

Preparation of the residue test sample

Place 1 g of the sample to be analysed in a 250 ml Erlenmeyer flask. Add 21 ml of HCl and 7 ml of HNO3, reduce to dryness, then add 30 ml of HCl + H2O. Boil, filter, transfer the filtrate to a 200 ml volumetric flask after cooling, and homogenise.

Calculation of the aliquot

The solution obtained must be transferred directly; if this does not give a result, a dilution must be made.

In atomic absorption, work in a hydrochloric acid medium, adding HC1 % (20% of the volume of the dilu).

Volume of the flask of choice / dilution = xml

The result X (%) obtained is calculated using the following formula:

$$X (\%) = \frac{(\text{Lecture courbe x100 x dilution})}{(\text{Prise d'essai x1000})} \quad (3.3)$$

(Curve reading x 1000 dilution)
(Test sample xlOOO)

Preparation of the test sample of a solution (filtered)

Calculation of the aliquot: proceed directly if necessary, otherwise dilute

In atomic absorption, work in a hydrochloric environment, adding HC1 % (20 % of the dilution volume)

X (mg/1 or g/1) = (curve reading x dilutions)

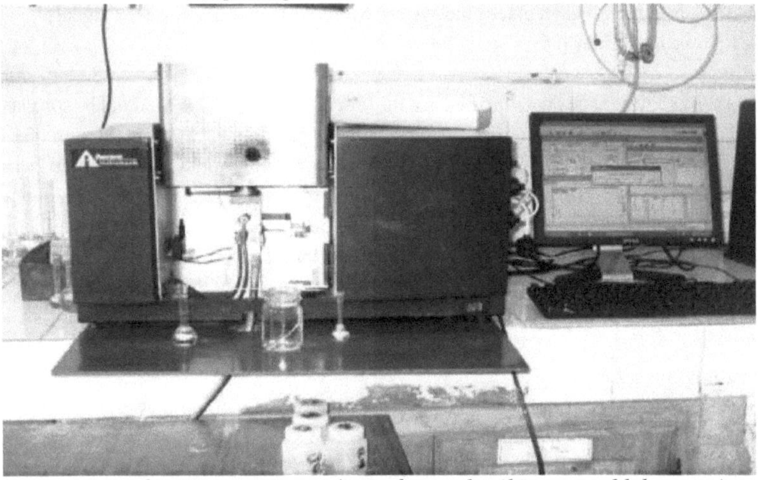

Figure 3.1: Atomic absorption apparatus (according to the Alzine central laboratory)

3.2.4. Fourier transform spectrometers (FTIR)

Fourier transform infrared spectrometers, which appeared around 1970, have a simple single-beam setup that differs from sequential devices in that the monochromator is replaced by an interferometer, often of the Michelson type, placed between the source and the sample. The radiation from the source strikes a *splitter,* consisting of a germanium film deposited on a KBr plate. The semi-transparency of this device allows two beams to be generated, one of which is directed towards a fixed mirror and the other towards a movable mirror whose distance from the splitter can be varied. These two beams, which are then recombined along the same path, pass through the sample before striking the detector, which measures the overall light intensity received.

The key component of the Michelson interferometer is the moving mirror, the only moving part, which oscillates over time between two extreme positions. When its position is such that the paths taken by the two beams are of equal length, the composition of the light leaving the interferometer is identical to that of the incoming light. However, when the moving mirror leaves this particular position, the light has a variable composition due to the phas t between the two paths: the signal transmitted over time by the detector is translated into an *interferogram,* I = f(5), where δ represents the difference in optical path length between the two paths (*Figure 3.2).*

The optical bench is managed and data acquired by a specific electronic interface. While the mirror is moving, an ADC converter connected to the detector samples the interferogram in the form of thousands of measurement points. Each of these values corresponds formally to the second member of a formidable linear equation whose unknowns are the amplitudes of all the wavelengths (assumed to be finite in number) present in the signal to be analysed. From these thousands of values, a specialised microprocessor calculates a giant matrix in less than no time at all, following a specific fast Fourier transform algorithm, to obtain the amplitudes of each wavelength in the spectral band under study. Taking into account a resolution factor imposed by the calculation method, we obtain the classic representation of the spectrum I = f(X) or l = f (v$^{(-)}$)**[80-83].**

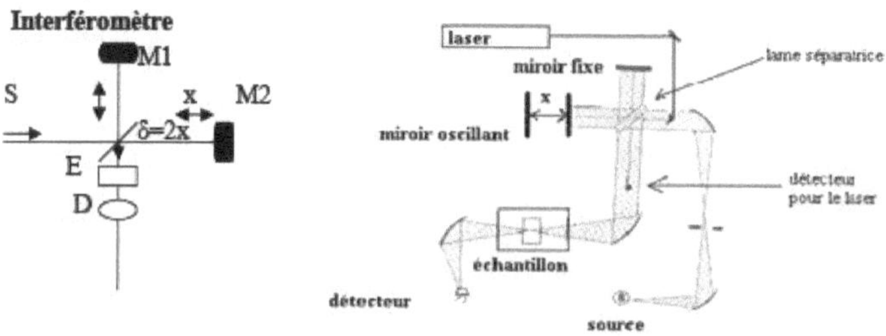

Figure 3.2: Optical setup of a Fourier transform device

The infrared measurements were performed on a JASCO model 4200 type A Fourier transform spectrometer. The samples were prepared in the form of 13 mm diameter pellets consisting of 2 mg of product diluted in 200 mg of KBr. The results are presented as transmittance for wavelengths between 4000 cm$^{'1}$ and 400 cm$^{'1}$.

3.2.5. X-ray diffraction

a/ Principle of X-ray diffraction

Incident X-rays sent through a crystal lattice are reflected by the planes of atoms that compose it. The interference of coherent waves diffused by each atom follows the direction given by Bragg's law:

$$n \lambda = 2 d_{hkl} \sin\theta \quad (3.4)$$

Where λ is the wavelength of the incident radiation, Θ is the angle between the incident radiation and the diffracted plane (hkl), dhki is the distance between two planes (hkl) and n is a positive integer proportional factor **[84,85].**

Apparatus

This technique is one of the most commonly used to characterise the various crystalline phases present in a compound. The method consists of sending a beam of X-rays with a well-defined wavelength λ, of the same order of magnitude as the interatomic distances, onto the solid sample. The X-rays are then scattered by the atoms. If these atoms are arranged in a crystalline structure, the scattered rays interfere with each other, causing the phenomenon of diffraction. The measurements were carried out using a Bruker D5005 diffractometer with Θ-Θ geometry at the Physical Measurements Department of the UMR 6503 research unit at the University of Poitiers (France). This consists of a copper anode (λκαΐ= 1.5406 Å) and cobalt (λκαΐ= 1.789 Å) (for the identification of the pigment "haematite") and a rear graphite monochromator that allows the elimination of Kß radiation and any fluorescence.

Sample preparation consists of grinding the sample to obtain a fine powder. This powder is then deposited on a sample holder or on a pure silicon wafer in the case of small masses (5 mg to 20 mg of sample). The assembly is then held in the centre of the diffractometer chamber by a magnet. The sample holder rotates around a vertical axis so that the crystallites are randomly oriented and no diffraction direction is favoured. The analysis conditions can be modified (X-ray incidence angle range, step and acquisition time) in order to refine the diffractograms obtained.

The different crystalline phases present in the sample are identified by comparing the experimental diffractograms with the PDF (Powder Diffraction File) files of the ICDD (International Centre for Diffraction Data) **[84, 86, 87]**.

3.2.6. Inductively coupled plasma atomic emission spectrometry

a) Optical emission spectrometry of atoms

When a chemical element in its atomic state is subjected to suitable excitation conditions, it emits radiation that is characteristic of that element. This observation forms the basis of a very general form of qualitative and quantitative elemental analysis, also known as OES (Optical Emission Spectroscopy). The optical emission of atoms is complex. It produces thousands of rays accompanied by a continuous background. An atomic emission spectrometer consists of (*Figure 3.3*):

> A device for introducing and dissociating the sample in the form of atoms or ions;

> An optical system for separating the different optical emissions;

> A system for detecting and analysing emitted radiation

> A computer system to interface with the user.

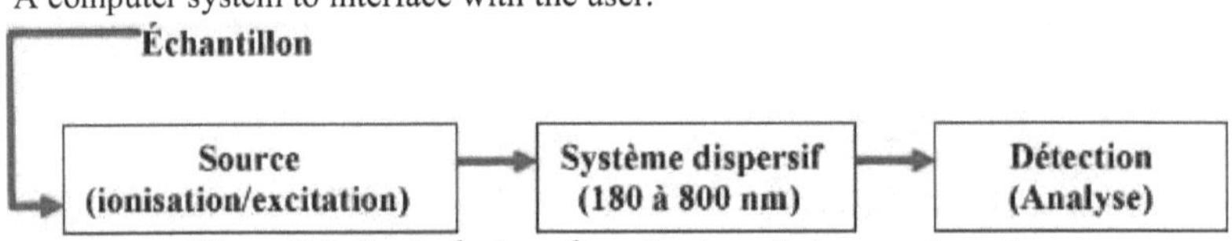

Figure 3.3: Basic design of an atomic emission spectrometer

To transform the sample into excited atoms or ions, processes based on gas plasmas, sparks, lasers or luminescent discharges are used. In our case, we used an inductively coupled plasma atomic emission spectrometer. The measurements were carried out on an OPTIMUM 2000 DV device in the Physical Measurements Department of UMR 6503 at the University of Poitiers (France) **[88-90]**.

b) Excitation by inductively coupled plasma torch

The principle of inductively coupled plasma emission spectrometry is based on the formation of plasma in a rare gas flow. Inductively coupled plasma is formed from an electrical discharge created in a flow of argon gas circulating through a series of concentric quartz tubes (torch) surrounded by a coil (induction coil). This is connected to a radio frequency generator. The passage of an alternating current through the induction coil produces an electromagnetic field that generates induced currents. As the torch is ly supplied with argon, a spark is produced to excite the argon gas. The electrons are accelerated by the electromagnetic field, resulting in a collision between argon atoms and the subsequent production of a larger number of electrons and argon ions, which are themselves accelerated. This process continues until the gas is highly ionised. The end result is the formation of a plasma with a temperature of 10,000 degrees Kelvin. The liquid sample is nebulised and then transmitted to the plasma. It undergoes various stages of decomposition, atomisation and ionisation, leading to the excitation of atoms and ions. After excitation, the atoms contained in the sample emit light with a characteristic wavelength. The light is transmitted through the optical system (network + prism) to a CID detector (video camera), which enables measurement. The detection limits for trace elements vary depending on the element: they are in the order of ppb or tens of ppb **[88-90].**

3.3. Laboratory tests

To better understand the problem of leaching residue recovery, it is essential to carry out various types of residue analysis. These types of analysis make it possible to identify the different components and their concentrations in the residues. Tests were carried out at the ALZINC central laboratory on solid leaching residues using chemical and spectral methods. The aim of this study is to extract zinc mainly from residues with a zinc content exceeding 20%. Parallel analyses carried out by ALZINC showed zinc contents in the leaching residues in the form of Zn (tot), Zn(H2O) and Zn(H2SO4), which are due to neutral and acid solution as a result of the influence of iron in the concentrate.

Tableau 1 1: Analysis of roasted zinc concentrate and leaching residue according to ALZINC (four batches of zinc concentrate) [44]

Periods	Cwide				Residue			
	Zn	Fe	$S(SO_4^{2-})$ %	S(S") %	Zn(total) %	$Zn(H_2O)$ %	$Zn(H_2SO_4)$ %	Fe
1	67.2	2.55	2.65	0.35	16.56	5.24	11.24	14.50
2	63.89	4.16	2.75	0.40	17.14	4.42	7.99	17.60
3	62.89	5.50	2.90	0.32	18.37	5.51	11.16	22.90
4	60.49	8.15	2.35	0.21	20.86	5.51	12.36	24.50
5	61.05	9.14	2.45	0.16	23.28	9.54	15.63	26.65

Tableau 2 2: Chemical composition of leaching residue according to ALZINC

Percentage content	01/2008	02
Zn	20.6	2
$Zn(H_2SO_4)$	9.5	9.9
$Zn(H_2O)$	6	6
Cd	0.16	0.14
Fe	26.35	25.38
P	4	4
Cu	0.82	0.85
Mn	0.44	0.63
CI	0.014	0.014
F	0.06	0.05
Mg	0.77	0.73
Co	0.0057	0.007
As	0.05	0.05
Ni	0.0036	0.0038
Ca	2.86	1.33
Ge	0.013	0.013
Cr	0.008	0.0069
Tl	0.001	0.001

These results show that for an iron-rich zinc load, the dissolution for zinc extraction by leaching is reduced, which means that the zinc content in the solid residues from leaching waste has increased.

This motivated our work in this project to find a better zinc extraction process when the concentrate load contains iron levels above 4% and to increase zinc extraction efficiency.

3.3.1 Chemical dosages

After taking samples of the solid residues, we carried out two types of analysis:

4- Zinc assay

4- Iron assay

4 Acid-base titration (acidity of the RC solution)

3.3.1.1 Zinc analysis

The zinc content in the residue was measured in terms of total zinc, as was the total zinc content in the filtrate. The results are shown in the tables.

a) Zinc determination in residues

a.1) Total zinc determination in residues

Procedure: In a 600 ml Erlenmeyer flask

Introduce 1.25 g of the residue to be analysed, attack with 21 ml HCl and 7 ml HNO3, dry thoroughly. Resume with 10 ml HCl + 10 ml H2O + 5 ml H2SO4, boil. Transfer everything to a 250 ml graduated flask already containing 10 ml H2O2 + 50 ml NH4CI, neutralise with NH4 OH + 15 ml in excess, cool and then measure to 250 ml with cold H2O, filter. Take 50 ml of the filtrate in a beaker, add 1 drop of bromothymol blue, neutralise with HC1 1/1 until the solution turns yellow. Add 200 ml H2O, plus 30 ml pH 5.5 buffer solution and 4 drops of orange xylenol. Titrate with E. D. T. A until the solution turns yellow.

Calculation:

$$\text{Zinc \%} = \text{Titrant volume x Titre}$$

b) Determination of total zinc in solutions

Take 1 ml of the solution using a pipette, add 200 ml of H2O, 30 ml of the buffer solution and a few drops of orange xylenol. The red colour of the zinc-dithizone complex will appear.

Titrate with EDTA until the red colour disappears and is replaced by a greenish-yellow colour, indicating the formation of the zinc-EDTA complex, which will destroy the zinc-dithizone complex.

Calculation:

$$\text{Zn (g/l)} = \text{Titrant volume x Titre}$$

Advantages of the buffer solution

It is best to work at a pH that is neither too acidic, so that the EDTA has good complexing power, nor too basic, to avoid unwanted reactions such as the formation of the Zn(OH$^{)}{}_{2}$complex. A buffer pH of 4 to 5 is a compromise that allows good titration of Zn^{2+}.

3.3.1.2 Iron titration

a) Iron determination in the residue

Procedure: in an Erlenmeyer flask

Attack 1 g of sample with 7 ml HNO3 and 21 ml HCl. Evaporate to dryness. Add 10 ml HCl and 20 ml HClO4, evaporate to white fumes. Allow to cool. Add 50 ml H2O and 10 ml H2O2. Neutralise with NH$_{(4)}$ OH, filter while hot in a 500 ml Erlenmeyer flask, then return the precipitate to the 500 ml Erlenmeyer flask. Wash once with HC1 1/1, then three times with hot water.

Evaporate the solution to 200 ml, add SnC12 drop by drop until the iron is completely reduced (the solution becomes colourless), allow to cool, then add 200 ml H2O + 15 ml HgCu + 10 ml H2SO$_4$ at 20% + 5 ml H3PO$_4$ + 3 drops of phenylamine. Titrate with K2CГ2O7.

Calculation:

Iron (g/1) = Titrant volume x Title

b) Determination of Fe^{++} e in solution

Procedure

Take 10 ml with a pipette into a 100 ml flask, then transfer to a 50 ml beaker. Add 25 ml HC1 + 10 ml Hr8O$_4$ at 20% + 5 ml H3PO$_4$. Add 3 drops of phenylamine

Titrate with IVCpO until the solution turns blue-violet

Calculation

Fe^{++} (g/1) = Titrant volume x Titre

3.3.1.3 Acid-base titration (acidity of the RC solution)

The acidity of the RC solution was determined by acid-base titration using colour indicators (methyl orange).

Procedure

Using a pipette, take 5 ml of the solution and place it in an Erlenmeyer flask.

Add 200 ml of H2O and a few drops of H2O2, as well as 3 drops of methyl orange Titrate with 1.02 N NaOH drop by drop until the colour changes.

3.4. Methods for extracting zinc from leaching residues

Insoluble residues containing lead and silver, such as PbSO4, AgCl, aluminates and silicates, are concentrated at a rate of 10 times that of the roasted ore. However, the resulting solution contains high levels of iron and impurities. It would therefore be useful to find a leaching process that dissolves as much zinc as possible in order to increase zinc extraction yield while avoiding the dissolution of iron and other impurities, or at least preventing them from entering the ZnSO4 electrolytic solution.

The methodology we used consisted of taking samples of the solid residues from the leaching plant effluent and carrying out laboratory tests using appropriate extraction methods. These tests were carried out under conditions very similar to those found in industry and to the leaching reactions at Sexploitation.

In the acid attack method, we chose sulphuric acid solution from electrolysis (return cells "RC"). The test consists of attacking a quantity of residue with the attack solution in a beaker, while studying the extraction variables, with agitation using a stirrer on a Bunsen burner. After the reaction time has elapsed, filtration is carried out to separate the solution from the sludge.

Approximately 10 kg of solid leaching residue was taken and 10 litres of sulphuric acid solution from electrolysis (return cells "RC") were added.

The laboratory equipment used was: Bunsen burner, pH meter, pH paper, thermometer, analytical balance, rotary stirrers and laboratory glassware.

Before carrying out the treatment tests, a complete analysis of the chemical composition of the leaching residue to be treated and the spent electrolysis solution (cell return "RC") was performed.

Tableau 3 3: Chemical composition of the leaching residue to be treated

Zn	Zn(H_2O)%	Zn(H_2SO_4)%	Fe	Cd	Cu	Pb	Ag	As	H_2O%
21.25	7.16	10.3	26.37	0.10	0.59	3.05	0.02	0.0014	40.9

Tableau 4 4: Chemical composition of the etching solution (Cell return)

Zn(g/1)	Cd(mg/1)	Cu(mg/1)	Fe(mg/1)	Pb (mg/1)	Mn (g/1)	Mg (g/1)	H^+ (g/l)
47.54	0.34	0.14	5.26	1.76	2.69	4.90	162

3.4.1 Zinc extraction methods

3.4.1.1 Acid leaching method

The test consists of treating a quantity of residue (200 g) with a prepared acid solution (RC+ H_2SO_4) in a beaker, while varying the extraction parameters (reaction temperature, pH of the treatment solution, reaction time) and stirring with a stirrer on a hot plate. After the set reaction time, filtration is carried out to separate the solution from the sludge. By analysing the sludge and the solutions obtained, the effectiveness of the experiments can be assessed.

a) Extraction parameters

The effectiveness of the dissolution of leaching residues by acid attack was assessed by the extraction of zinc in solution. The concentration of sulphuric acid was set at 160, 200, 240 and 280 g/1, the leaching temperature at 85 and 90°C, and the leaching time at 120 and 180 minutes.

Procedure

The test consists of attacking a quantity of residue (200 g) with 1000 ml of a cell return solution at different acidities (160, 200, 240, 280 g/1) in a 2000 ml beaker and stirring with a stirrer on a hot plate. After the set reaction time, filter to separate the solution from the sludge. The effectiveness of the experiments can be assessed by analysing the sludge and the solutions obtained. Distilled water was added during the reaction to prevent continuous evaporation.

- Take 1000 ml of the prepared attack solution and 83 ml of distilled water, and place in a 2000 ml beaker.
- Leave on a hot plate for 5 minutes, then gradually add 200 g of residue while stirring.
- Check the temperature, trying to stabilise it (85 or 90 °C), as well as the degree of agitation.
- After the set time (120 or 180 minutes) for the reaction, proceed to filtration, filter and wash with approximately 20 ml of distilled water.

The residue obtained is sampled to prepare a significant sample for analysis: Leave the residue obtained in an oven for 2 hours at a temperature of 105°C, then grind the residue using a cylinder mill.

The results obtained from the residue and filtration analyses are reported in the following tables :

Table 3.5: Zinc content in wet extraction of zinc at acidity = 160 g/l

(Test sample of solid residue leaching m=200 g)

Time (minutes)		12	18	12	180
Temperature (°C)		85	85	90	90
Zinc content in residue (%)		**11.92**	**12.62**	**9.4**	**10.7**
Zinc content in the filtrate (g/i)	Zn (total)	**64.20**	**55.68**	**85.70**	**84.10**
	Fe (total)	**35.5**	**34.29**	**68.1**	**65.7**

Table 3.6: Zinc content in wet extraction of zinc at acidity = 200 g/l

(Sample of solid residue after leaching, m = 200 g)

Time (minutes)		12	18	12	180
Temperature (°C)		85	85	90	90
Zinc content in residue (%)		**9.70**	**10.24**	**8.92**	**9.90**
Zinc content in the filtrate (g/i)	Zn (total)	**59.9**	**52.6**	**79.9**	**64.2**
	Fe (total)	**41.3**	**39.2**	**56.7**	**44.1**

Table 3.7: Zinc content in wet extraction of zinc at acidity = 240 g/l

(Sample of solid residue after leaching, m = 200 g)

Time (minutes)		12	18	12	180
Temperature (°C)		85	85	90	90
Zinc content in residue (%)		**8.6**	**9.4**	**6.71**	**7.9**
Zinc content in the filtrate (g/i)	Zn (total)	**76.7**	**69.85**	**98.34**	**89.16**
	Fe (total)	**50.4**	**46.58**	**79.26**	**65.3**

Table 3.8: Zinc content in wet zinc extraction at acidity = 280 g/l

(Test sample of solid residue leaching m=200 g)

Time (minutes)		12	18	12	180
Temperature (°C)		85	85	90	90
Zinc content in residue (%)		**13.2**	**16.1**	**7.94**	**11.5**
Zinc content in the filtrate (g/i)	Zn (total)	**41.12**	**39.6**	**90.6**	**49.93**
	Fe (total)	**27.1**	**22.3**	**68.2**	**29.7**

3.4.1.2 Goethite method

3.4.1.2.1 Principle

This process consists of hot acid treatment of the leaching residues to precipitate the iron in the form of semi-hydrated iron hydroxide FeO.OH or $Pe_2\ O_3.H_2\ O$, which is easily decantable and filterable. To do this, the solution must contain a low ferric iron content, which brings us back to the following treatment:

To obtain the precipitation of *goethite,* the treated solution must always be maintained at a low ferric iron concentration, which leads us to do the following:

4- Preliminary reduction of ferric iron

Ά- oxidation and simultaneous precipitation of iron.

The reactions are as follows:

$$Fe_2(SO_4)_3 + ZnS = 2\ FeSO_4 + ZnSO_4 + S \quad (3.6)$$

$$2FeSO_4 + 2ZnO\ 1/2\ O_2 + H_2O = 2ZnSO_4 + Fe_2O_3.\ H_2O \quad (3.7)$$

Only one molecule of Zn is therefore needed to precipitate one molecule of Fe **[91]**. The process operations are shown in *Figure 3.7.*

a) Double hot acid leaching

The "leaching residue" is subjected to subsequent leaching at atmospheric pressure and a temperature of 90°C, using a dilute sulphuric acid solution with an initial free acid concentration of 240 g/l in an amount equal to 2 *to* 5 times the stoichiometric amount of acid required to dissolve the zinc contained in the leaching residue, which forms a sludge known as lead-containing residue, comprising recoverable insoluble elements such asthe acid required to dissolve the zinc contained in the leaching residue, forming a slurry hereinafter referred to as lead-bearing residue comprising recoverable insoluble elements such as lead, silver and gold. The filtrate from the first hot acid leaching will be sent to the reduction stage. The filtrate from the second hot acid leaching will be sent to the head of the neutral leaching **[91]**.

b) Iron reduction

This second operation consists of reducing ferric iron to ferrous iron by adding a reducing agent such as a sulphide ore, comprising in particular zinc sulphide for economic reasons, and also to avoid introducing foreign elements into the solution that could be harmful to the overall process, including purification and the electrolysis room. This reduction is carried out according to the following reaction:

$$2\,Fe^{3+} + ZnS \longrightarrow Zn^{2+} + 2\,Fe^{2+} + S \quad (3.8)$$

The operation takes place at a temperature between 90 and 95°C and lasts approximately 4 hours. Compared to the theoretical quantity required, an excess of 15 to 20% ZnS is added, which is why, after settling of the mixture, the pulp containing the excess raw ore is then filtered and the insoluble material is sent to the roasting stage **[91]**.

c) Neutralisation

The solution is neutralised after reduction by adding the required amount of calcine to maintain a pH of between 1.7 and 3.5, so that the iron precipitates, taking with it other impurities such as arsenic, lead and fluorine. The reaction time is 4 hours at a temperature of 90°C.

After neutralisation, decantation produces a residue which is sent to the first hot acid leaching stage **[91]**.

d) Oxidation. Hydrolysis of iron

The removal of iron dissolved in the previous stages is the specific operation of wet treatment of residues; it must also remove impurities redissolved in large quantities during the leaching of ferrites. The solution resulting from neutralisation is oxidised and the iron it contains is precipitated using coram, an oxidising agent, and oxygen from the air, which is necessary for the formation of iron oxide at a temperature of 90°C for a period of 4 hours. The reaction for precipitating iron using the Goethite process is as follows:

$$2\,FeSO_4 + ZnO + 1/2O_2 + H_2O = ZnSO_4 + Fe_2O_3 + H_2O + 2\,FeOOH \quad (3.9)$$

In one form of the process, oxidation and precipitation are applied to a solution with a density ranging from 1.20 to 1.5 and an iron content of no more than 50 g/l. Oxidation is accelerated in the presence of small amounts of anions such as copper, which has a catalytic effect and complexes with ferric ions. The solution must contain more than 300 mg/1 of copper **[91,92]**. The following diagram shows the steps in the Goethite process:

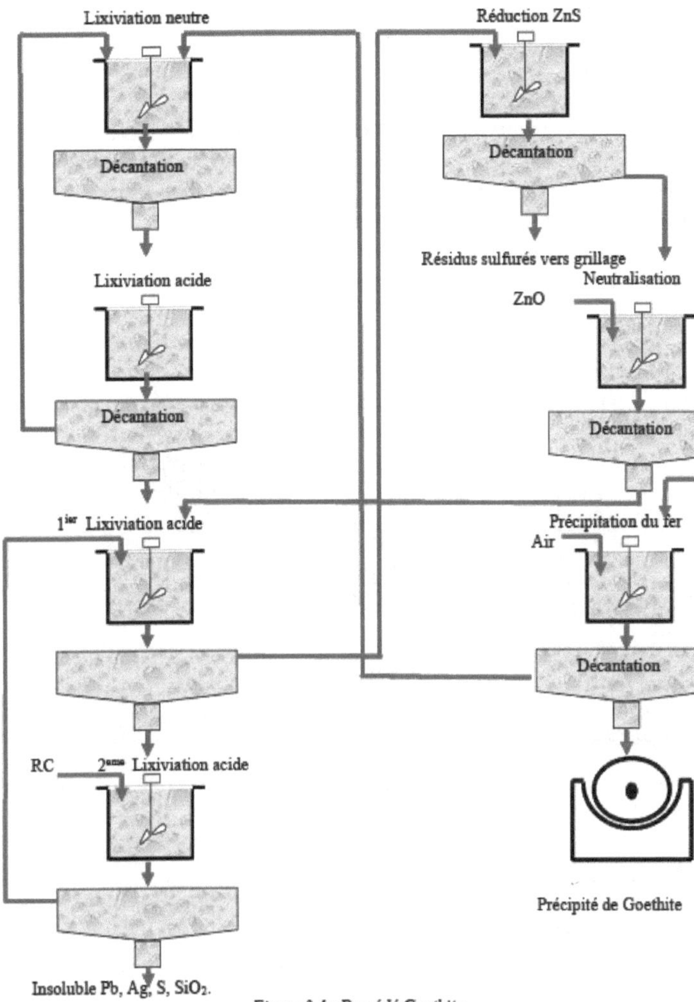

Figure 3.4 : Procédé Goethite

Figure 3.4: Goethite process

3.4.1.2.2 Synthesis of Goethite

I^er^phase

The experiment consists of attacking a quantity of residue (200 g) with one litre of an aqueous solution of sulphuric acid at 240 g/l H2SO4, prepared in advance from the acid returned from the tanks in which the electrolysis of zinc sulphate and concentrated acid is carried out.

Procedure

1ère stage of the experiment

In a 2000 ml beaker, add 1000 ml of a cell return solution prepared at 240 g/l H2SO4 and 100 ml of H2O.

- Stir using a Bunsen burner for heating. When the temperature reaches 90°C, add 200 g of residue little by little while stirring.

• Check the temperature at 90°C, trying to stabilise it, as well as the degree of agitation. After the set time (2 hours) for the reaction, proceed to filtration, washing the filter with about 20 ml of distilled water.

2nd stage of the experiment

The filtrate is sent to the reduction phase, the residue in solution is washed and dried and undergoes a second hot acid leaching under the same conditions as the first solution. After the set time (2 hours), a lead-containing residue is separated by filtration, and the solution obtained is returned to the head of the first hot acid leaching.

• The residues obtained are sampled to prepare significant samples for analysis: leave the residues obtained in an oven for 2 hours at a temperature of 105°C, grind the residues using a cylinder mill.

The various elements were measured in the charge products on the one hand and in the solution residue and lead-containing solutions on the other.

The analyses of Zn, Fe and acidity were carried out by volumetric analysis, while Pb, Cu, Cd, Ag and As were measured by atomic absorption. The results are shown in Table 3.9 below:

Table 3.9: Analyses of the different products I^er^phase

elements	Leaching residue %	Return Cell g/i	Solution after dissolution g/1	Residue after dissolution %	Final solution g/i	Lead residue %
Zn (total)	21.2	4	91	8.6	35.26	4.74
Fe (total)	26.37	0.003	58.50	5.68	6.01	2.72
Pb	3.05	0.00152	0.00013	4.45	0.00163	9.81
Cd	0.10	0.00003	0.0005	0.03	0.01647	0.02
Cu	0.59	0.00002	1.523	0.16	0.1225	0.02
Ag	0.02		0.000091	0.0205	0.00013	0.041
As	0.014		0.04605	0.001	0.00896	0.016
H2SO4	-	131	125	-	129	-

2nd phase

The filter from the first phase undergoes a reduction of ferric iron to ferrous iron through the addition of raw blende.

Procedure

In a 2000 ml beaker, add 1000 ml of the solution filtrate and 100 ml of H2O.

• Let it stir with a stirrer using a Bunsen burner for heating. When the temperature reaches 95°C, add 28 g of ZnS little by little while stirring.

• Control the temperature between 95 and 100°C, trying to stabilise it, as well as the degree of stirring.

After the set reaction time (4 hours), proceed to filtration, washing the filter with about 20 ml of distilled water.

After reduction, the pulp undergoes solid-liquid separation, from which a sulphurous residue is collected and sent for roasting.

• The sulphurous residue and the filtrate obtained are sampled: the residue is left in an oven for 2 hours at a temperature of 105°C, then ground using a roller mill. The analyses of certain elements are recorded in Table **3.10:**

Table 3.10: Analysis of the different products of the second phase

Elements	Solution before reduction g/l	Raw blend %	Solution after reduction g/l	Sulphur residue %
Zn (total)	91.8	51.10	81.8	7.3
S (total)	0.5032	33.25	0.41159	36.73

Fe (total)	58.5	7.61	54.30	9.09
Fe^{2+}	0.15	-	38.54	-
Pb	0.00013	0.64	0.00056	0.04
Cd	0.0005	0.13	0.01179	0.015
Cu	1.523	0.28	1.241	0.06
As	0.0046	0.021	0.00968	0.0041
H2SO4	125	-	128	-

3rd phase

This phase consists of lowering the acidity of the filtrate after reduction using calcine as a neutralising agent. The pH must be maintained between 1.7 and 3.5, and the particle size of the calcine must be less than 74 μm.

Operating procedure

Test 1

In a 2000 ml beaker, add 1000 ml of the filtrate from the previous phase and 100 ml of H2O.

- Stir using a stirrer with a Bunsen burner for heating. When the temperature reaches 90°C, add the required amount of ZnO little by little while stirring until the pH is between 1.7 and 2.5.
- Control the temperature at 90°C, trying to stabilise it, as well as the degree of agitation.
- 300 mg of copper sulphate was added to act as a catalyst.

After the set reaction time (4 hours), filtration is carried out and the filter is washed with about 20 ml of distilled water.

After neutralisation, the pulp is filtered, and the residue is collected and sent for the first hot acid leaching.

- The residue and filtrate obtained are sampled: the residue is left in an oven for 2 hours at a temperature of 105°C, then ground using a roller mill.

Test 2

In a 2000 ml beaker, add 1000 ml of the filtrate from the previous phase and 100 ml of H2O.

- Stir using a stirrer with a Bunsen burner for heating. When the temperature reaches 90°C, add the required amount of ZnO little by little while stirring until the pH is between 2.5 and 3.5.
- Control the temperature at 90°C, trying to stabilise it, as well as the degree of agitation.
- Add 300 mg of copper sulphate to act as a catalyst.

After the set reaction time (4 hours), filtration is carried out and the filter is washed with about 20 ml of distilled water.

After neutralisation, the pulp is filtered, and the residue is collected and sent to the first hot acid leaching process for further processing.

- The residue and filtrate obtained are sampled: the residue is left in an oven for 2 hours at a temperature of 105°C, then ground using a roller mill.

The analyses of certain elements of the starting products of this operation and of the products obtained are shown in Table **3.11:**

Tableau 11 1: analysis of the different products of the third phase (neutralisation)

elements	Solution before neutralisation g/1	Calcined blend ZNO%	Solution after neutralisation test 1 g/1	Residue after neutralisation test 1 %	Solution after neutralisation test 2 g/1	Residue after neutralisation test 2 %
Zn (total)	81.8	59.	136.	2	157.60	28.12

Total Fe)	54.30	8.18	9.22	9.32	25.10	20.90
Fe^{2+}	38.54		3.66		15.46	-
Pb	0.00056	0.86	0.00033	0.94	0.00094	0.88
Cd	0.0118	0.15	0.1845	0.33	0.15	0.452
Cu	1.241g/l	0.44	1.170	1.82	1.628	1.642
As	0.00968	0.014	-	0.024	0.00081	-
H2SO4	128	-	0	-	0	-

4th phase

The final step consists of oxidising the previous filter with oxygen from the air. Oxidation and precipitation are applied to a solution with a density ranging from 1.20 to 1.5. with a maximum iron content of 50 g/l and a copper content of around 250 mg/l, which is sufficient to double the reaction speed.

Procedure

Test 1

In a 2000 ml beaker, add 1000 ml of the filtrate after neutralisation and 100 ml of H2O.

- Stir using a stirrer with a Bunsen burner for heating. When the temperature reaches 90°C, blow in enough air to oxidise the iron.
- Check the temperature at 90°C, trying to stabilise it, as well as the degree of agitation.

After the set reaction time (4 hours), proceed to filtration, washing the filter with about 20 mi of distilled water.

After oxidation, the pulp is filtered, and a dark yellow precipitate called "Goethite" is collected. The final solution containing zinc in solution is returned to the top of the neutral leaching process.

- The residue and filtrate obtained are sampled: the residue is left in an oven for 2 hours at a temperature of 105°C, then ground using a roller mill.

Test 2

In a 2000 ml beaker, add 1000 ml of the filtrate after neutralisation and 100 ml of H2O.

- Stir using a stirrer with a Bunsen burner for heating. When the temperature reaches 90°C, blow in sufficient air to oxidise the iron.
- Check the temperature at 90°C, trying to stabilise it, as well as the degree of agitation.

After the set reaction time (4 hours), proceed to filtration, washing the filter with about 20 ml of distilled water.

After oxidation, the pulp is filtered, and a red precipitate called "Goethite" is collected. The final solution containing zinc in solution is returned to the top of the neutral leaching process.

- The residue and filtrate obtained are sampled: the residue is left in an oven for 2 hours at a temperature of 105°C, then ground using a roller mill.

Table **3.12** shows the analyses of the different elements before and after oxidation.

Tableau 12 2: analyses of the different products of the 4th oxidation phase

elements	Solution before oxidation 1 g/i	Solution before oxidation 2 g/i	Solution after oxidation 1 g/i	Solution after oxidation 2 g/i	Iron precipitate 1	Iron precipitate 2
Zn (total)	136.7	157.6	72.7	95.2	6.67	1.87
Fe (total)	9.22	25.10	1.94	0.919	28.80	48.22
Fe^{2+}	3.66	15.46	-	-	-	-
Pb	0.00033	0.00094		0.00077	0.02	0.
Cd	0.1845	0.15		0.136	0.086	0.02
Cu	1.17	1.628		0.774	0.32	0
As		0.00081		0.00042		0.0084

Ag	0	-	-	0.000089	-	0.01

3.4.2.1.3 Inductively coupled plasma emission spectrometry

Table 3.13: Analysis of the different elements present in Goethite precipitates (tests 1 and 2)

Analyte	Goethite % (Test 1)	Goethite % (Test 2)
Zn	6.349	1.7
Fe	30.36	42.50
P	0.062	0.053
Cr	0.053	0.025
Cu	0.143	0.325
C	0	0
Co	0.002	0.002

3.5. Transformation of Goethite - Hematite

3.5.1. Principle

This work concerns a process for manufacturing an iron oxide-based pigment, specifically from a sulphate solution containing ferrous sulphate obtained from the treatment of zinc leaching residues.

To recover certain elements, particularly zinc. Zinc metallurgy residues can be treated using various processes, including a process in which the residues are subjected to subsequent leaching using sulphuric acid, which produces, among other things, a solution of zinc sulphate and ferric sulphate. which is reduced by the addition of zinc sulphate to ferrous sulphate, and the iron is removed in the form of oxides in a residue that constitutes a waste product with no commercial value or use. Our work consists of finding a use for this residue by transforming it in a simple and economical way into an iron-based pigment, which is a product of great industrial and commercial value with applications in many fields **[93]**.

The following figure shows the principle of the hematite process:

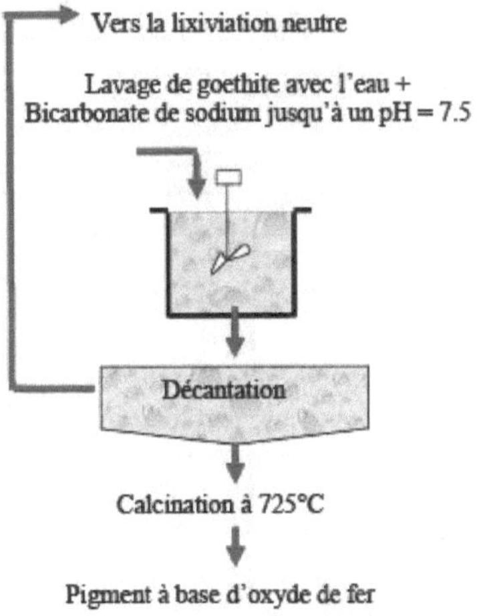

Towards neutral leaching

Washing goethite with water +
Sodium bicarbonate until pH = 7.5
Calcination at 725°C
Iron oxide-based pigment

Figure 3.5: Hematite process

5.2. Hematite synthesis

The Goethite precipitate obtained from zinc hydrometallurgical leaching residues is first washed in a basic medium at a temperature of 90°C for 6 hours, then with ordinary water and calcined at 725°C for 4 hours, thus transforming it into a red pigment called "haematite" which has the desired pigment characteristics [**93**].

Procedure

In a 1000 ml beaker, add 2 g of goethite and 800 ml of H2O.

- Stir with a stirrer using a Bunsen burner for heating. When the temperature reaches 90°C, add enough baking soda to raise the pH to 7.5.
- Control the temperature at 90°C, trying to stabilise it, as well as the degree of agitation. After the set time (6 hours) for the reaction, proceed to filtration, washing the filter with about 20 ml of distilled water.
- After washing in a basic medium, the pulp is filtered and the precipitate is calcined for 4 hours at 725°C, with the following values:

Table 3.14: Analysis of the different elements before and after transformation of goethite to hematite

Elements	Goethite %	Hematite %
Total Zn	1.8	1.35
Fe (total)	45.3	58.4
Pb	0.01	0.09
Cd	0.02	0.0039
Cu	0.029	0.07
As	0.008	-
Ag	0.01	0.001
Mn	0.008	0.04
Ni	0.0085	0.021
Co	0.005	0.011
Mg	0.034	0.19
Cr	0.007	0.023
K	0.008	0.137
Ca	0.039	0.166

Conclusion

In the first part of this chapter, we presented the characterisation methods used, their principles, as well as the various laboratory tests and chemical analyses. We then carried out various types of residue and cell return analyses, which enable us to identify the different constituents and their concentrations.

The main objective of this study is to recover zinc from leaching residues in zinc hydrometallurgy, where the problem is its dissolution. Given that the double leaching process only achieved a rate of 90%, the project involves acid leaching using the RC cell return solution and a sulphuric acid solution reconstituted at 240 g/1, followed by precipitation of the iron in the form of goethite, an iron oxide that can be recovered as a dye, for example, or for other uses.

The methodology we used consists of taking samples of the solid residues from the leaching plant effluent and carrying out laboratory tests using a double extraction method:

- First acid extraction using the sulphuric acid solution from the electrolysis cells as the attack solution. We conducted tests at different temperatures and pH levels, modifying the reaction kinetics by varying the reaction time. The zinc recovery rates were 12 to 14% and 98.34 g/1 in residues and solution, respectively, with the parasitic dissolution of undesirable impurities, particularly iron. The results showed that zinc extraction in solution depends on many parameters, including the nature of the chemicals present in the solid residue, the pH of the reaction medium, the temperature and the reaction time. The optimal extraction values are an acidic medium (240 g/1), sufficient time for complete redissolution (120 minutes) and a temperature between 90 and 95 °C.
- Extraction by double leaching followed by the Goethite process phases using a prepared sulphuric acid solution (return from electrolysis cells + H2SO4) as the attack solution, taking into account the physical and chemical conditions of the reactions, with maximum zinc recoveries of 14.75% and 72.70 g/1 in Goethite (test 1) and 19.75% and 72.70 g/1 in Goethite (test 2). - chemical conditions of the reactions, with maximum zinc recoveries of 14.75% and 72.70 g/1 in Goethite (test 1) and 19.39% and 95.21 g/l in Goethite (test 2) in residues and solution respectively, with no risk of iron and impurities dissolving. According to this method, iron is eliminated by forming an insoluble complex, the concentrations of which in the goethite precipitates (tests 1 and 2) are 29.58% and 45.69%, 1.94 g/l and 0.919 g/l in the solution that returns to the head of the neutral leaching, respectively.
- The recovery of goethite precipitate (test 2) into an iron oxide-based pigment yields a red precipitate with a zinc content of 1.352% and 58.40 % iron.

In the next chapter, we will interpret the results of quantitative analyses (atomic absorption spectrometry and inductively coupled plasma atomic emission spectrometry) and present the various results of qualitative analyses (Fourier transform spectrometers (FTIR) and X-ray diffraction).

CHAPTER 4

Results and discussion

Introduction

The management of special industrial waste (SIW), such as hydrometallurgical leachates, is one of the major challenges of sustainable development and the environment. In the context of increasingly stringent environmental regulations, it is necessary to find an environmentally friendly, effective and inexpensive technique for managing these residues. In zinc production by hydrometallurgy, leachate residues are estimated at over 90 tonnes/day for a daily output of 300 tonnes [**94**]. These sludges are known as zinc ferrite sludges. Laboratory analyses have enabled us to identify the forms of zinc/пOPerO3 and determine the chemical composition (Table 3.9), with zinc and iron contents estimated at 21.25% and 26.37% respectively. Similar work [**95**] confirmed our results, with zinc ferrites diffracting at **20**= 35.2°, 42.9°, 53°, 56.5°, 62°, 69° (Figure 4.1).

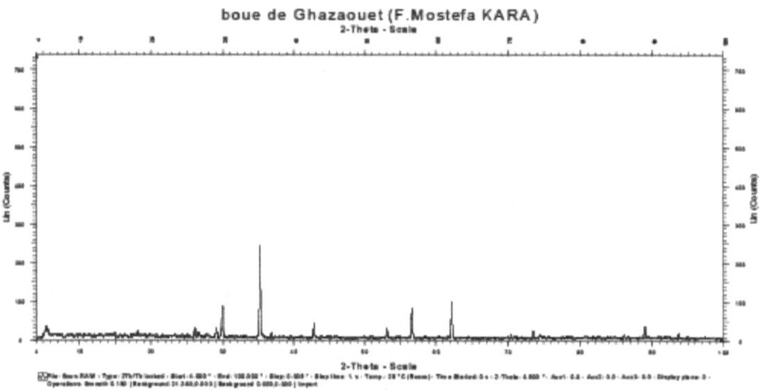

Figure 4.1: X-ray diffractogram of Uxiviation residues [**95**]

Another chemical species has been identified, such as anglesite (PbSO4), where **20** = 29.9° and 42.9°.

Among the solutions proposed for the management of D.I.S. waste from hydrometallurgy, we have open-air storage, isolated landfill storage, and treatment and recovery processes. In this study, we are interested in the recovery of zinc through the physical-chemical treatment of leachate residues. Given the diversity of elements contained in leachate sludge, we will focus on zinc recovery. To do this, we proceeded as follows:

- Acid extraction using RC sulphuric acid solution from the return flow of the electrolysis cells as the attack solution. We will carry out tests at different temperatures and pH levels, modifying the reaction kinetics by varying the time.
- Double leaching extraction followed by the Goethite process using a prepared sulphuric acid solution (return from electrolysis cells + H2SO4) as the attack solution, taking into account the physical and chemical conditions of the reactions.
- Goethite precipitate will be recovered by transforming it in a simple and economical way into an iron oxide-based pigment, which is a product of great industrial and commercial value with many uses.

4.1 Influence of leaching parameters

Leaching parameters such as temperature, reaction time, pH, reaction medium, chemical composition of the ore, etc. influence the composition of the leach residues. The leaching yield depends on several parameters, but in zinc extraction, the electrolyte must be highly pure, as the maximum amount of zinc must be extracted without putting other elements into solution, as these elements reduce the electrolysis yield and deteriorate the quality of the zinc obtained.

The main factor influencing the dissociation of zinc during leaching is the presence of iron in the concentrates. When a blende is rich in iron (more than 6%), which is often the case, ferrites, mixed complexes of zinc and iron, are formed during roasting according to the following reaction:

$$ZnO + 2\,FeS + 7/2\,O_2 \longrightarrow ZnOFe_2O_3 + 2\,SO_2 \quad (4.1)$$

These ferrites are insoluble in the dilute sulphuric acid used in hydrometallurgical operations and their presence hinders these operations. Zinc forms insoluble complexes with iron.

If the leaching temperature decreases, this can lead to a drop in zinc yield, and if it increases, this does not necessarily lead to a good yield, as undesirable elements can pass into solution and interfere with zinc extraction. The temperature must be between 55 and 60°C.

The leaching reaction time may be insufficient if the flow rates increase. This means that the dissolution of ZnO will not be sufficient and the zinc yield will decrease. The optimal reaction time is 30 minutes in each mixing tank.

The acidity of the cell return solution (RC solution) must be maintained at approximately 160 g/L. A decrease in acidity will result in a decrease in zinc extraction. The purity of the electrolysis solution must not contain any impurities in solution. These elements reduce the electrolysis yield and the quality of the zinc obtained.

The leaching medium must be oxidising in order to convert ferrous iron into ferric iron, but the presence of sulphur sulphide, which is the case with poor roasting and poor reduction quality, leads to a decrease in yield and the dissolution of impurities.

The agitation speed can improve the dissolution kinetics, but the installations are designed for a flow rate of 50 m^3/h. If the speed is increased, there is a risk of reducing the leaching yield and the dissolution will be poor.

Acid methods for extracting zinc from leaching residues and precipitation methods will be presented in the following paragraphs:

4.2 Acid methods

The acid extraction method for zinc from solid leaching residues was developed by several researchers, A. Bendjebbour and A. Benmoussat **[44]**, who used RC cell return solution (158 g/l acidity) and other extraction parameters, varying the pH of the attack solution at constant temperature and time in the range 1.5–3.5; variation in the temperature of the attack solution at constant pH and reaction time in the range 55°C - 75°C; variation in reaction time while maintaining the temperature and pH of the attack solution constant. The time is equal to 90 and 120 minutes.

The results showed that zinc extraction in solution depends on many parameters, including the nature of the chemicals present in the solid residue, the pH of the reaction medium, the temperature, the reaction time and the extraction method. The zinc initially lost in the leaching residues can be recovered by acid extraction with an average content of 3-9% in the residues using the RC solution with a pH adjusted to 2.5, a temperature below 70°C and a sufficiently long time t = 120 minutes.

In our study, we carried out acid extraction using an adjusted solution, varying the sulphuric acid concentrations from 160 g/1 to 280 g/1, while the temperature was maintained between 85 and 90°C and the leaching time was between 120 and 180 minutes.

4.2.1 Influence of acidity

By varying the acidity of the acid solution composed of RC solution and sulphuric acid, we obtained maximum dissolution at 240 g/l. The results showed good solubility of zinc in the solution, which can reach 98.34 g/l for a concentration of 240 g/l of sulphuric acid, corresponding to 79.26 g/l of iron. Similarly, we observed very high dissolution of zinc ferrite and an increase in zinc in solution with increasing concentrations of sulphuric acid used and lower zinc losses in the residues. According to the curves (Figures 4.2 and 4.3), for concentrations below 240 g/l (160-240 g/l), extraction increases, reaches a maximum and then decreases for concentrations between 240 and 280 g/l.

The variations in acidity enabled us to determine a maximum value that will be used to recover zinc from zinc ferrite residues. This value is 240 g/l.

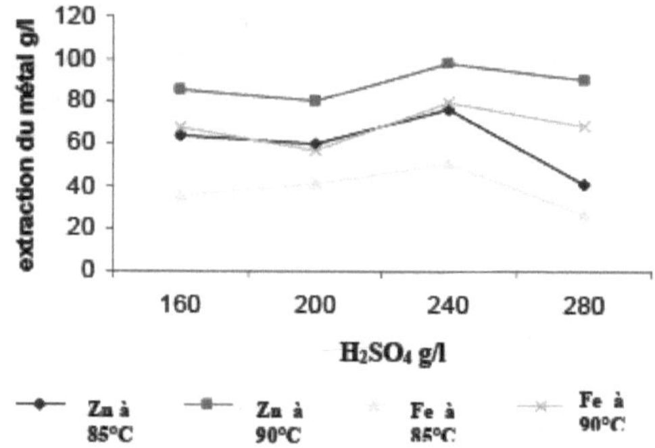

Figure 4.2: Zinc and iron content (g/l) as a function of acidity at 120 min

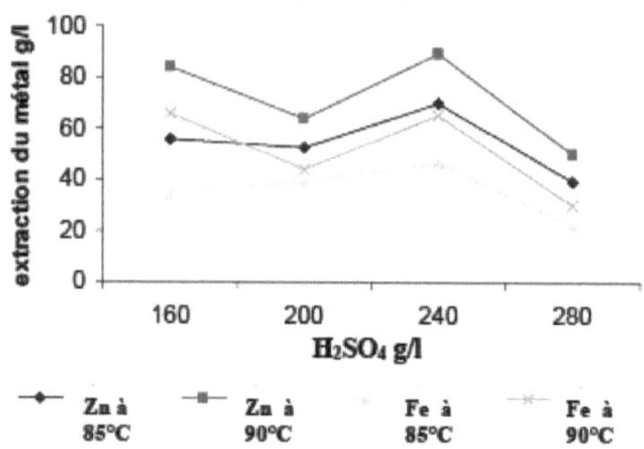

Figure 4.3: Zinc and iron content (g/l) as a function of acidity at 180 min

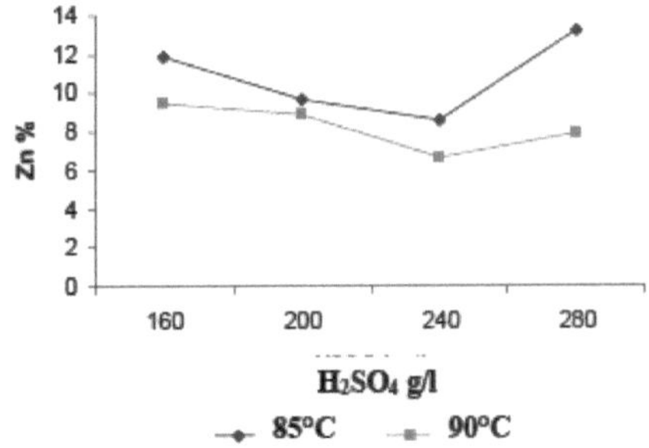

Figure 4.4: Zinc content % as a function of acidity at 120 min

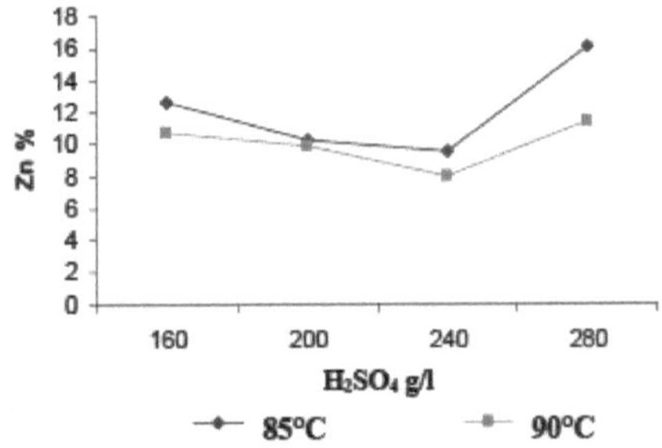

Figure 4.5: Zinc content (%) as a function of acidity at 180 minutes

Tests varying the immersion time from 120 minutes to 180 minutes, for acidity values ranging from 160 to 280 g/1, showed that increasing the time beyond 120 minutes is unnecessary, given the losses in zinc in solid form as the time increases (Figures 4.4 and 4.5).

4.2.2 Influence of temperature

Tests varying the temperature between 85 and 90°C showed that the zinc content in the residue is low as the temperature increases (Figures 4.6 to 4.9). Maximum zinc concentrations in the filtrate, 98.34 g/1, were obtained at a maximum temperature of 90°C for an immersion time of 120 minutes. Tests at the same temperature and with an extended time of 180 minutes showed a decrease in zinc concentrations in the ZnSO4 solution due to the redissolution of undesirable elements.

An increase in the temperature of the solution during the attack considerably increases the zinc recovery yield (Figures 4.6 - 4.9). It can be said that temperature plays a decisive role in zinc extraction. An increase in temperature during dissolution allows the zinc to be extracted in the filtrate and reduces zinc losses in solid form in the residues.

The solubility of zinc and iron increases with rising temperatures, hence the best extraction yield during the four acidity tests was at 90°C, 120 minutes and 240 g/1 of sulphuric acid.

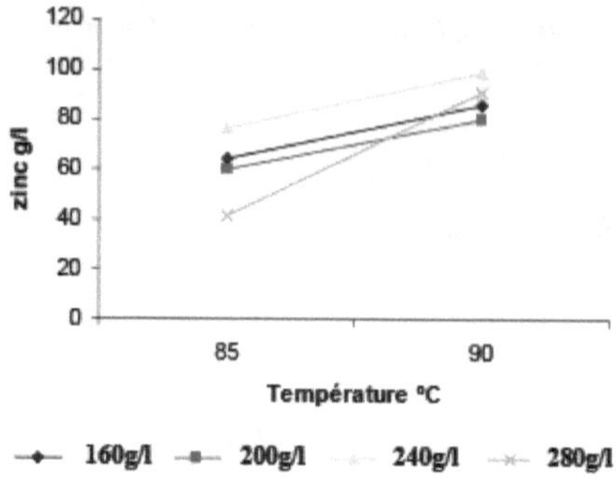

Figure 4.6: Zinc content (g/l) as a function of temperature at 120 minutes

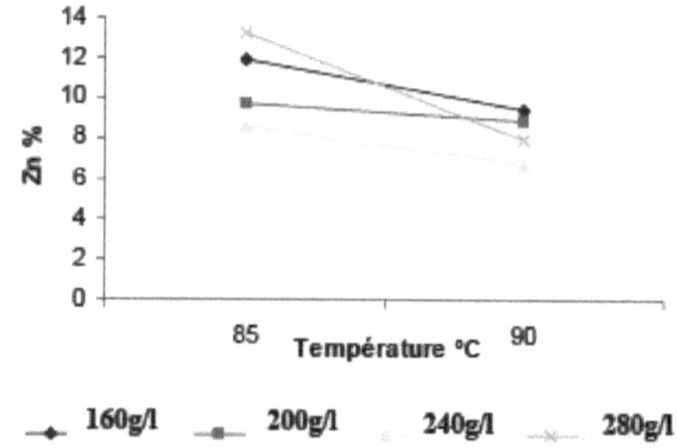

Figure 4.7: Zinc content (%) as a function of temperature at 120 minutes

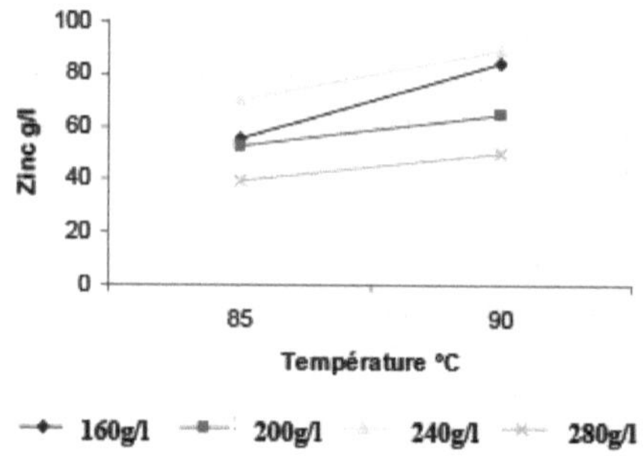

Figure 4.8: Zinc content (g/l) as a function of temperature at 180 min

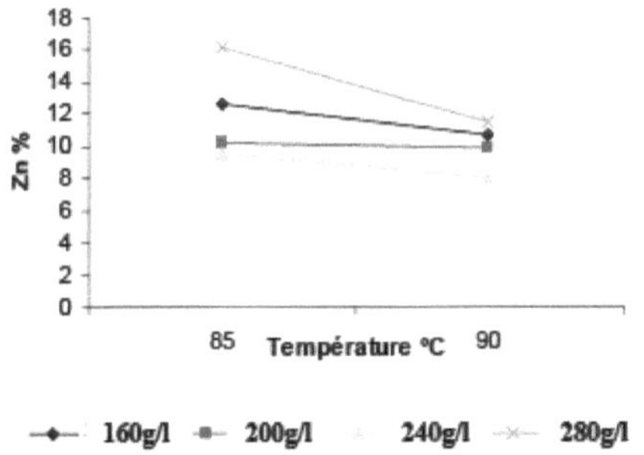

Figure 4.9: Zinc content % as a function of temperature at 180 min

4.2.3 Influence of immersion time

Tests to recover zinc using the acid method by varying the immersion time in the interval (120–180 min) and the temperature (85–90°C) showed that increasing the immersion time reduces the zinc yield regardless of the temperature value in the selected interval (Figures 4.10–4.13).

A zinc content of 98.34 g/l in the filtrate was obtained for an acidity of 240 g/l at a temperature of 90°C and an immersion time of 120 minutes. The zinc content in the residues decreased to 14%, representing a recovery rate of 65.88%.

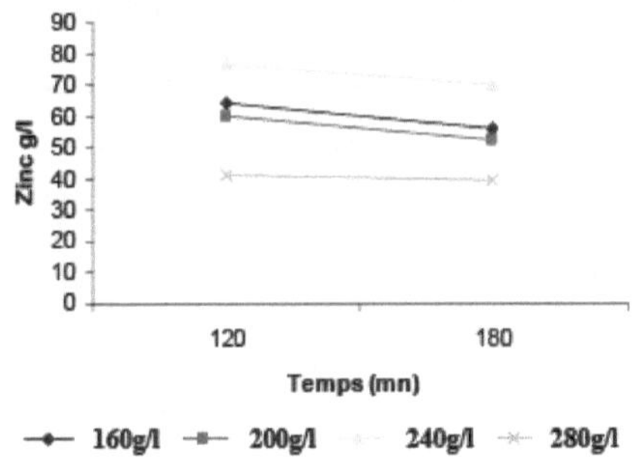

Figure 4.10: Zinc content (g/l) as a function of time at 85°C

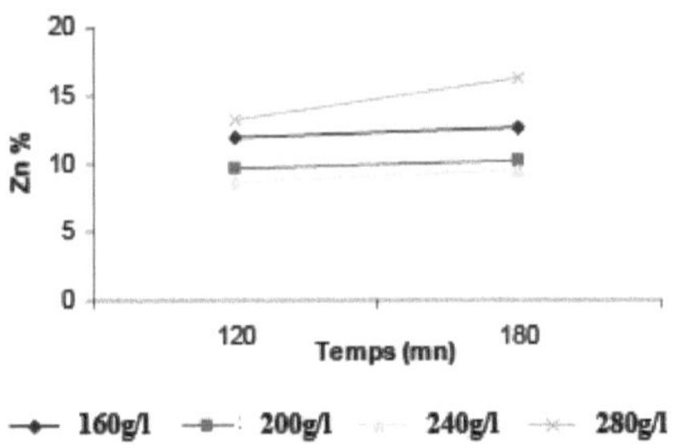

Figure 4.11: Zinc content (%) as a function of time at 90°C

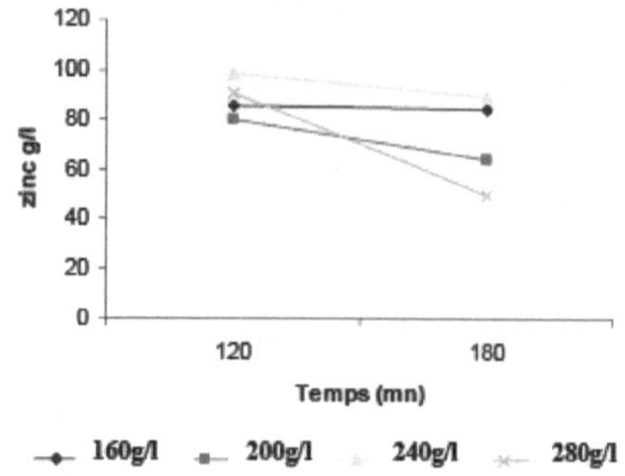

Figure 4.12: Zinc content (g/l) as a function of time at 90°C

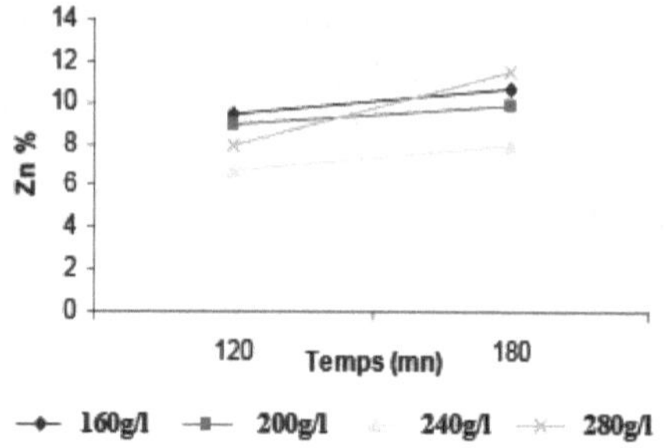

Figure 4.13: Zinc content (%) as a function of time at 90°C

Conclusion

In view of these results, despite variations in the extraction parameters, the zinc extraction rate from zinc ferrite residues, estimated at 65.88%, remains insufficient due to the presence of iron and undesirable elements that pass into solution. We have developed a complementary process involving iron precipitation, which removes all impurities before zinc electrolysis in the form of an iron-based complex compound called "Goethite", an easily filterable iron hydroxide.

4.3 Goethite precipitation method

Zinc extraction using the acid process remains insufficient, with quantities of zinc remaining in the sludge. We have therefore developed a complementary process called "Goethite", which consists of precipitating iron and undesirable elements in the form of an insoluble complex contained in the ZnSO4 solution.

4.3.1 Synthesis of Goethite

The Goethite process **[91]** aims to dissolve zinc ferrites, which are insoluble complexes in leaching. After acid leaching, it includes a phase of reducing ferric iron to ferrous iron, a neutralisation step and a final phase of oxidation and precipitation in the form of goethite.

- *Reduction phase*

The reduction was carried out using raw ZnS ore as a reducing agent. The reduction took place at 95°C for 4 hours according to the following reaction:

$$Fe_2(SO_4)_3 + ZnS \longrightarrow ZnSO_4 + S + 2\,FeSO_4 \quad (4.2)$$

After filtration, a sulphurous residue is obtained, which is returned to the roasting process.

The gearbox was chosen so as not to introduce any new elements into the double leaching zinc refining process.

- *Neutralisation phase*

The solution will be neutralised with calcine to a pH of between 1.7 and 3.5 at 90°C for 4 hours. The undissolved calcine will be added to the head of the hot acid leaching.

- *Oxidation and precipitation phase*

The filtrate produced by neutralisation is heated to a temperature of 90°C and oxidised by adding air for 4 hours.

Iron precipitation is achieved in less than one hour, but stirring is continued for several hours in order to dissolve as much of the added calcine as possible.

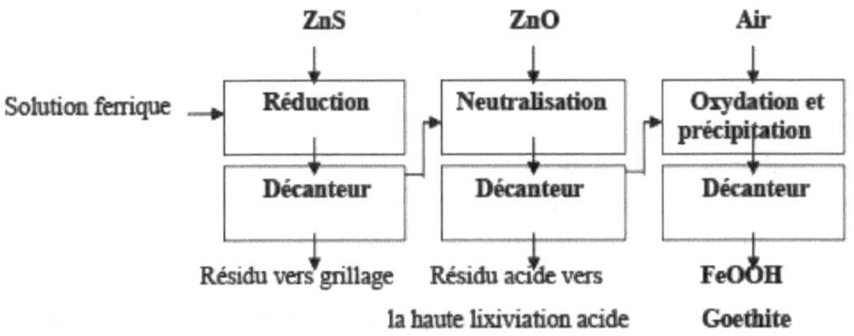

Figure 4.14: Simplified diagram of the Goethite process

4.3.1.1 Influence of zinc solubility

The first dissolution test consists of taking a quantity of residue (200 g) and dissolving it in one litre of an aqueous solution of H2SO4 at 240 g/1, prepared in advance from the RC solution and concentrated sulphuric acid. The results are presented in Table 3.9. According to these results, the solubility of zinc also affects the solubility of iron, with maximum concentrations in the filtrates of 91.8 g/1 and 58.5 g/1, respectively, which resulted in a significant decrease in zinc and iron contents in the residues, 8.62% and 5.68% respectively, i.e. a residue recovery rate of 59.63% in the first solution.

The quantities of undesirable elements such as lead, cadmium and silver decrease in solution and increase in residues. A slight increase in copper and arsenic is observed, which may have been trapped in zinc ferrites in solution, and their decrease in residues.

The second dissolution test consists of a hot acid attack under the same operating conditions and with the sa . The results are shown in Table 3.9. They show that the zinc extraction rate has improved from 8.62% in the first dissolution to 4.74% in the residue treatment, with a rate of 35.26 g/l in solution.

Iron, copper, cadmium, silver and arsenic are present in low concentrations in the solution and residue.

Analyses of the final residue showed that it contains a recoverable lead content of 9.81% (Table 3.9), known as lead residue, which will be the subject of further research.

4.3.1.2 Reduction of iron III to iron II

This test was carried out in order to transform the form of iron with the aim of eliminating it by oxidation in the final phase of "Goethite". The results (Table 3.10) of the reduction phase of ferric iron to ferrous iron indicate a slight decrease of 10 g/1 of zinc in solution and 7.37% in the residue. There is also a slight increase in lead, cadmium and arsenic in the residue and in solution due to the raw blende used for reduction.

With regard to iron, there was a minimal decrease of 4.2 g/l of total iron in solution and an

increase from 0.15 to 38.54 g/1 in ferrous iron.
The reduction yield of ferric iron to ferrous iron was 70.97%.

4.3.1.3 Neutralisation

The purpose of this test is to adjust the pH of the solution to a value at which iron precipitates. Calcined zinc oxide (ZnO) was used as a neutralising agent in the pH ranges 1.7 to 2.5 and 2.5 to 3.5 with a particle size of less than 74 µm. The neutralising agent was chosen with the aim of maintaining the same chemical species in the zinc sulphate solution. The results (Table 3.11) show that the solubility of zinc in solution is 157.60 and 136.70 g/1 for the first and second pH ranges, respectively, and that of iron is 9.22 g/1 and 15.46 g/1. High zinc levels of 27.14% and 28.12% were obtained for both pH ranges, due to poor dissolution of the calcine as a result of low pH values.
Undesirable elements such as lead, arsenic, cadmium, copper and silver have variable values, decreasing for lead and increasing for the others in solution and in residue.

4.3.1.4 Iron oxidation and precipitation

The final step involves oxidising the previous filtrate with oxygen from the air. Oxidation and precipitation are applied to a solution with a density of 1.20 to 1.5, an iron content of 9.22 and 15.46 g/l and a copper content of around 250 mg/1 to double the reaction speed. Two precipitates were obtained according to the pH intervals and will be characterised in the following paragraph (atomic absorption, inductively coupled plasma emission spectrometry). The final solution containing zinc in solution will return to the neutral leaching phase. The results are shown in Table 3.12.
The results of the atomic absorption analyses gave the zinc and iron contents in the residue and solution for the first and second pH ranges: dark yellow precipitate, 6.67% and 72.7 g/1 zinc and 28.80% and 1.94 g/1 iron; red precipitate: 1.87% and 95.21 g/1 of zinc and 48.22% and 0.919 g/1 of iron. These results were confirmed by residue analysis using inductively coupled plasma emission spectrometry (Table 3.13), where the contents were: 6.34% and 1.77% for zinc and 30.36% and 42.50% for iron.
Comparison of the results for the two intervals shows that the zinc extraction yield is better at a pH between 2.5 and 3.5, i.e. 91.43%.
The iron precipitation phase is characterised by the removal of certain impurities, such as lead, cadmium, copper, arsenic and silver.
Figures 4.15 and 4.16 show the evolution of zinc and iron according to the Goethite treatment stages for the two pH ranges.

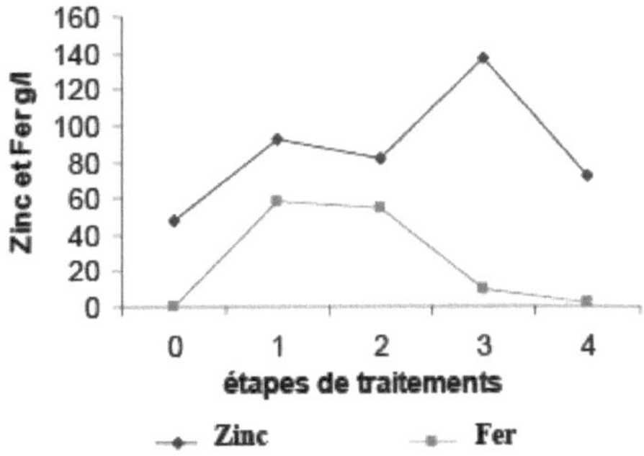

Figure 4.15: Evolution of zinc and iron (g/l) as a function of treatment stages in the pH range 1.7Ò2.5

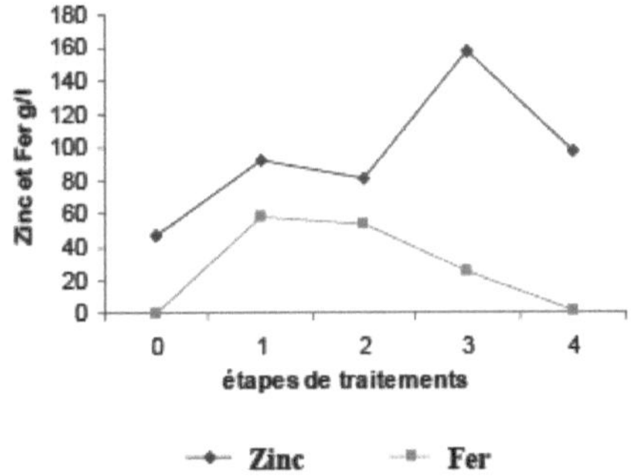

Figure 4.16: Evolution of zinc and iron (g/l) as a function of treatment stages in the pH range between 2.5 and 3.5

The results of precipitation extraction in the form of "Goethite" (Table 3.12) show that zinc concentration values in solution in the neutralisation phase are up to 157.60 g/1 and were obtained for a less acidic pH (2.5 to 3.5) and the iron concentration is around 25.10 g/l, an acceptable value below 50 g/l according to the goethite process. During the oxidation phase, a loss of zinc was observed. The new concentration is 95.21 g/l, which represents a further loss of zinc of 62.39 g/1, while the iron concentration has decreased to 0.919 g/l.

The new zinc loss can be explained by the trapping of zinc in the goethite precipitate obtained.

4.3.2 Characterisation of goethite

In our investigation at the zinc extraction laboratory using the Goethite process, we obtained zinc in solution in the form of ZnSO4, which will be extracted by electrolytic deposition and two solid precipitates according to the pH range (1.7 to 2.5) and (2.5 to 3.5) of dark yellow and red colour, which will be characterised in the laboratory by taking isolated samples for microanalysis by X-ray diffraction and Fourier transform infrared spectroscopy.

4.3.2.1 X-ray diffraction

The X-ray diffractograms were obtained according to the selected pH range values (Figures 4.17 and 4.18). They show that the main compounds are goethite and hematite.

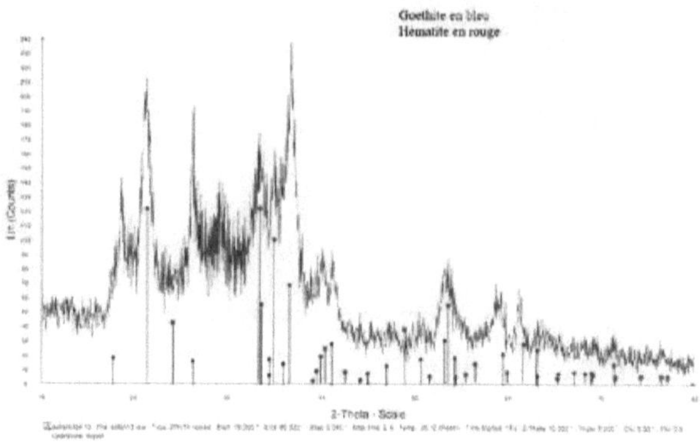

Figure 4.17: X-ray diffractogram (1.7> pH < 2.5)

Figure 4.18: X-ray diffractogram (2.5> pH < 3.5)

The different phases were identified by comparing the diffractogram obtained with the JCPDS (Joint Committee on Powder Diffraction Standards) charts (Appendix B). The Rx diffractogram (Figure 4.17) shows a partial crystallisation range between **20**= 18° and 42°, characteristic of a non-crystalline or amorphous phase, and a crystalline phase. The peaks representative of goethite for 1.7 > pH < 2.5 appear at **20**= 18.1°, 21.2°, 26.2°, 33.2°, 34.8°, 36.7°, 39.9°, 41.1°, 45°, 47°, 50.5(°) and 53° corresponding to Miller indices hkl respectively: [200], [101], [201], [301], [210], [111], [211], [401], [311], [410],[112]and [[212]. These values are comparable to those reported by M. Pelino [**96**]. It can be seen that, according to the X-ray diffractogram (Figure 4.17), the peaks are slightly broadened. This phenomenon may be caused by the presence of small crystallites (less than 200 nm) according to [97] and the amorphous phase.

Other phases have been identified in small proportions, with the most intense peaks at **20**= 24.1°, 33.1°, 35.6°, 40.8°, 54.1°, corresponding to Miller indices hkl: [012], [104], [110], [113], [024], [116] characteristic of hematite, as well as zinc in oxide form at **20**= 31.8°, 34.5°, 36.8° and 62.7° and in zinc ferrite form at **20**= 29.2°, 34.8° and 61.3°.

The X-ray diffraction pattern (Figure 4.18) shows a partial crystallisation range between **20=** 21° and 42°, characteristic of an amorphous phase, and the presence of a crystalline phase.
The representative peaks appear at **20=** 18°, 21.6°, 26.6°, 33.5°, 34.8°, 36.1°, 39.9° and 41.1°, corresponding to Miller indices hkl respectively: [200], [101], [201],[301], [210],[lll], [211], [401] and [212] characteristic of goethite. These values are comparable to those reported by the same author **[96].** These peaks are less broad compared to those obtained in Figure 4.17, where the proportion of goethite decreases.
Other phases were identified with varying proportions, with the most intense peaks at **20** = 24.1°, 33.1°, 35.6°, 40.9°, 49.5°, 54.1°, 62.5° and 64.1° with hkl values of [012], [104], [110], [113], [024], [116], [214] and [300], which are comparable to those given by M. Pelino **[96]** as characteristic of hematite. These results highlight the strong presence of hematite (TerO$_3$) as the major phase in the composition, due to the increase in pH, which favoured the transformation of goethite during the oxidation of ferrous iron to ferric iron during treatment, while trapping low proportions of zinc (1.82%) in the iron precipitate.
Another phase was identified at **20=** 31.8°, 34.5°, 36.8° and 62.7°, characteristic of oxide, and another at **20=** 29.5°, 35° and 63.3°, characteristic of zinc ferrite.

4.3.2.2 FTIR spectra

The precipitates obtained were characterised by Fourier transform infrared spectroscopy (FTIR) in two pH ranges (1.7 > pH < 2.5) and (2.5 > pH < 3.5). The spectra are shown in Figures 4.19 and 4.20. This technique highlights the various functions present in the solid compounds, where goethite and hematite are the main components and are identified by X-ray diffraction at different proportions.
The FTIR spectra show that in both pH ranges selected, the precipitated goethite has the same appearance and that the peak intensities are less intense for the first pH range.

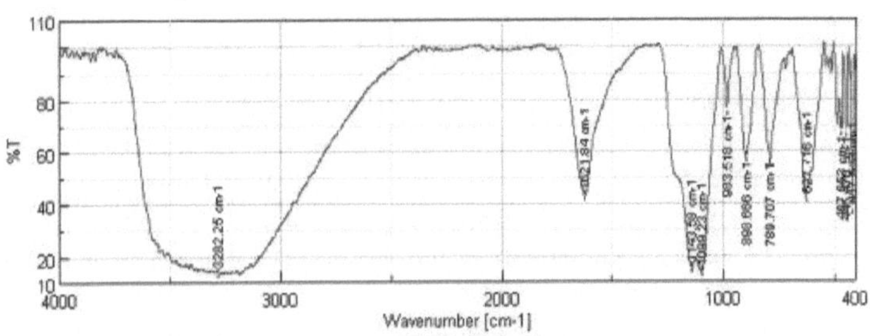

Figure 4.19: FTIR spectrum (Fourier transform infrared) (1.7> pH < 2.5)

The spectrum in Figure 4.19 shows:

- A very broad valence band at 3282.25 cm'^{1} from the hydroxyl groups in the leaflet (v-OH). The intensity and position of this band depend on the degree of hydration and the nature of the metal cations, respectively. It can be seen that this band contains two bands at 3400 and 3216.68 cm'($^{1)}$ (Figure 4.20). This result is consistent with the work of Rúan. H. D **[98];**
- a band of high intensity at 1143.58 cm"$^{(1)}$ 'which is an asymmetrical band called the O band. According to Claudio Morterra et al **[99],** this band is observed in all FTIR spectra characteristic of naturally occurring or synthetic goethites. Another band called the b band appears between 1100 and 900 cm'(1$^)$corresponding to the specific surface area of O"(2) 'In our case, this band appeared at 983.518 cm'(1) [99]**;**

- A deformation band outside the hydroxyl plane (γ-Fe-O-H) at 789.70 $^{cm^{-1}}$ **[100];**
- A deformation band in the plane (δ-OH) at 898.66 cm'^{1} and at 1621.84 cm'^{1} corresponding to the combination (γ-OH+ δ-OH) **[101];**
- at 627.716 cm'^{1} a stretching band (γ-Fe-O) as reported by Luiz C.A. Oliveira , Teodorico C. Ramalho, Eugênio F. Souzaa, Maraísa Gonçalvesa, Diana Q.L. Oliveira, Márcio C. Pereira and José D. Fabris [I02].

The values obtained for the bands corresponding to the characteristic absorption of pure goethite are 606 to 635 cm'^{1} , 858 to 890 cm'^{1,}1600 cm'l and 3100 to 3500 cm'(1) This difference is due to the purification rate of the synthesised goethite [**102**] **(**Appendix C).

- At 417.513 cm ^{1},a band of medium intensity corresponds to goethite according to Ruan et al. **[100].**
- Bands of medium intensity observed at around 467.653 and 445.476 cm'^{1} correspond to absorption bands characteristic of hematite **[100].**

The values obtained for the bands corresponding to the characteristic absorption of pure hematite are 475.20 and 533.58 cm'^{1} **[103];** this difference is due to the purification rate of the synthesized hematite (Appendix **C).**

The results obtained in the FTIR spectrum (Figure 4.20) show:

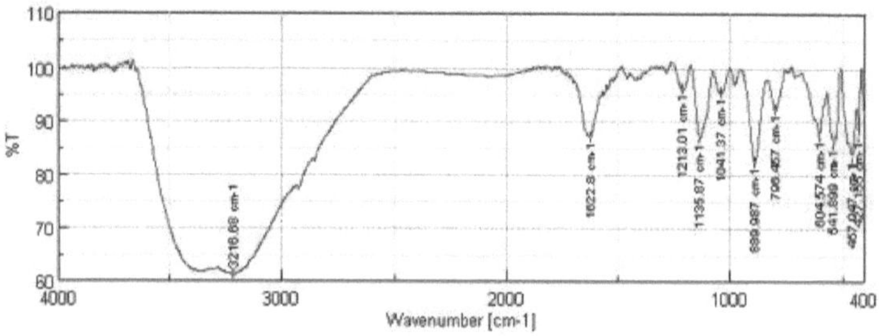

Figure 4.20: Fourier transform infrared spectrum (2.5>pH <3.5)

Several absorption bands corresponding to synthesised goethite and hematite:

- at 3400 and 3216.68 cm'^{1} two valence vibration bands of the hydroxyl groups of the sheet (v-OH). The intensity and position of this band depend respectively on the degree of hydration and the nature of the metal cations;
- a band of high intensity at 1135.87 cm'$^{(1)}$ This is an asymmetrical band called the O band. According to Claudio Morterra et al [**99**]**,** this band is observed in all FTIR spectra characteristic of naturally occurring or synthetic goethites. Another band called the a band appears between 1250 and 1350 cm"(1$^)$In our case, we have a band at 1213 cm"(1$^)$ which is characteristic of the specific surface area of oxygen [**99**]**.**
- At 796.45 cm'^{1} **,**a deformation band outside the hydroxyl plane (γ-OH) **[100] is observed.**
- a deformation in the piano (δ-OH) is observed at 889.98 cm'^{1} and at 1622.8 cm'^{1} resulting from the combination (γ-OH+ δ-OH) **[101];**
- at 604.476 cm'^{1} a valence band (γ-Fe-O) as reported by Verdonck et al. **[104],** Cambier also observed that the band around 630 cm'^{1} is affected by the shape of the a-FeOOH particles.
- at 541.899 and 457.047 cm'^{1}of medium intensity bands related to the vibration of (γ-Fe-O) of the Hematite compound, these results are in agreement with those in the literature by H.

Namduri, S. Nasrazadani **[103] and** at 427.155 cm'(1) band related to goethite according to Rúan et al. **[100].**
The FTIR spectrum results obtained for the synthesised goethite and hematite compounds were compared with those found in the literature by P.S.R. Prasad et al **[105],** Ruan et al. **[98],** Russell and Fraser **[106], and** Claudio Morterra et al **[99].**
The absorption band values obtained and those found in the literature are presented in Tables 4.1 and 4.2.

Tableau 1: : Absorption band values observed by Fourier transform infrared spectroscopy corresponding to the literature in the 400-4000 cm^{-1}region and Assignment

(a-P.S.R. Prasad et al, b- Ruan et al, Ruan et al, c- Russell and Fraser, d- Claudio Morterra et al)

a	b	c	d	Attribution
402	461	455,409	-	FeO6 lattice
640	619	633	-	FeO6 lattice
795	798	794	790	γ-deform-OH
887	888	893	890	òdeform-OH
-	-	-	800-1100	b-band
-	-	-	1140	O-band
-	1269	-	1250-1325	a-band
1634	1687,1643	-	1635	γbend-OH
3177	3206	3153	3140	v stretch(O-H)
3407	3450	-	-	v stretch (H-O-H)

Tableau 2: : Absorption band values observed by Fourier transform infrared spectroscopy corresponding to the synthesised compounds (goethite) in the 400-4000 cm^{-1}region and the Assignment

(Gì and G2 represent goethite synthesised according to pH intervals 1 and 2)

Gi	G2	Assignment
417.513	427.155	FeO6 lattice
627.716	604,574	FeO6 lattice
789.707	796.457	γ-deform-OH
898.666	889.987	δ- deform-OH
963.518	-	b-band
1143.58	1135.87	O-band
-	1213.01	a-band
1621.84	1622.8	γ- bend-OH
3282.25	3216.68	v-stretch (O-H)
-	-	v-stretch (H-O-H)

Based on these results, we can conclude that the synthesized solid compounds correspond to goethite and hematite with different proportions depending on the pH intervals. The proportion of goethite is high in the first pH interval ($1.7 > pH < 2.5$) compared to the second. This phenomenon can be explained by the fact that this pH range is favourable for the solid precipitation of goethite up to 2.5. Above this value, a colour change from yellow-brown to red was observed, which is characteristic of the precipitation of hematite, where the pH is favourable for its formation.

Work on these compounds in the literature confirms our results (Tables 4.1 and 4.2).

4.3.2.3 Microstructure of goethite

The phase structures identified are goethite and hematite. Scientific work **[99]** on the microstructure of these compounds has shown the type.

Goethite, which is an iron hydroxide with the chemical formula a-FeOOH, has an orthorhombic crystal structure with space group Pnma (No. 62) **[107-109].** The three-dimensional structure is built from $PeO_3(OH)_3$ octahedra forming tunnels that extend in the [010] direction and where the hydrogen atoms are located. Each iron octahedron is connected to eight neighbouring octahedrons by four edges and four vertices; the oxygen atoms are found in a tetrahedral environment OPe_3H or OFe3-H (strong H bond). The oxygen atoms define compact hexagonal layers he [**110-113].** The lattice parameters are: a = 9.95 Å, b = 3.01 Å, c = 4.62 Å.

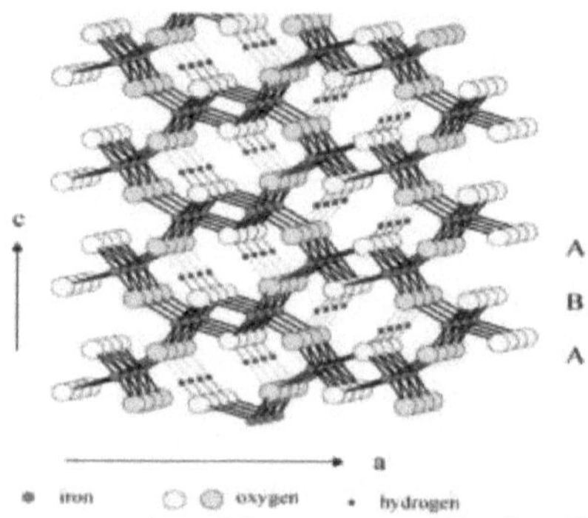

Figure 4.21: Schematic representation of the structure of Goethite **[99]**

4.4 Study on the recovery of goethite

The process of recovering zinc from goethite in leaching residues enabled us to recover zinc in solution at a rate of 91.43% and a precipitate in the form of oxides. This is a waste product with no commercial value.

We are interested in this precipitate in order to recover it as an iron-based pigment, which is a valuable product from an industrial and commercial point of view with applications in many fields.

The process involves transforming the synthesised goethite into hematite within the range (2.5> pH <3.5), where the proportions of hematite are significant according to the crystallographic characteristics ( ) and chemical composition (Zn =1.82% and Fe= 45.36%).

4.4.1 ***Synthesis of hematite***

The goethite precipitate obtained from zinc hydrometallurgical leaching residues is first washed in a basic medium at a temperature of 90°C for 6 hours, then with ordinary water and calcined at 725°C for 4 hours **[93],** This transforms it into a red pigment with the desired pigment characteristics, which has been characterised in the laboratory by X-ray diffraction and Fourier transform infrared spectroscopy.

4.4.2 ***Atomic absorption analysis***

The results in Table 3.14 showed that after washing in an alkaline medium and calcination, a precipitate was collected, and atomic absorption analysis of the solution after washing did not detect any elements.
There was also a slight solubility of zinc from 1.87 to 1.352%, as well as a significant increase in iron from 45.36 to 58.40%.
Similarly, there was a decrease in the quantities of undesirable elements such as lead, cadmium and silver. There was a slight increase in copper, manganese, nickel, cobalt, magnesium, chromium, potassium and calcium.

4.4.3 *Characteristics*

4.4.3.1 X-ray diffraction

The characterisation of the precipitate obtained after washing in an alkaline medium and calcination by X-ray diffraction (Figure 4.22) indicated that it has the crystallographic structures of hematite, confirming that a large part of the goethite was transformed into hematite during the calcination process.

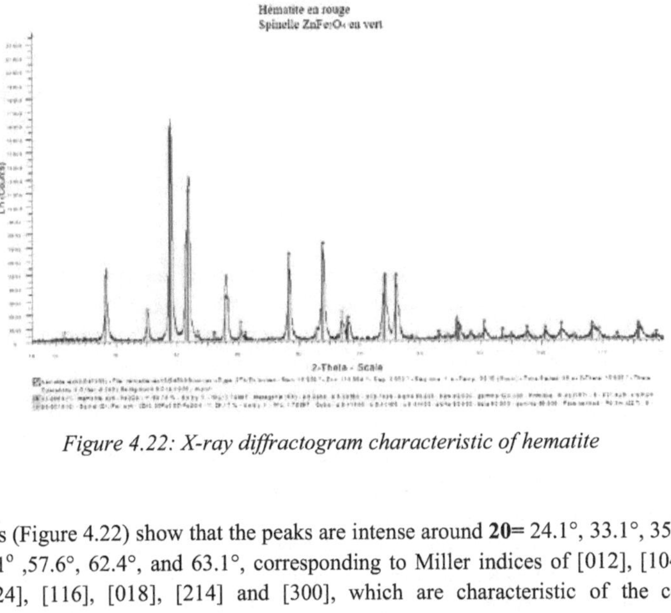

Figure 4.22: X-ray diffractogram characteristic of hematite

The results (Figure 4.22) show that the peaks are intense around **20=** 24.1°, 33.1°, 35.6°, 40.8° ,49.9°,54.1° ,57.6°, 62.4°, and 63.1°, corresponding to Miller indices of [012], [104], [110], [113], [024], [116], [018], [214] and [300], which are characteristic of the compound hematite.
Other peaks were found at 20=21°, 25°, 41.5°, 50.2°, 67°, and 74°, corresponding to the presence of the spinel phase ZnFe2O4. According to Thomas Verdier et al. **[114]**, this phase can form during the grinding of a mixture of zinc oxide ZnO and hematite α-ΓerO$_3$, so we assume in our case that this phase formed during the preparation of our sample (goethite *(2.5 >pH <.3.5))* during qualitative and quantitative analyses.

4.4.3.2 FTIR spectra

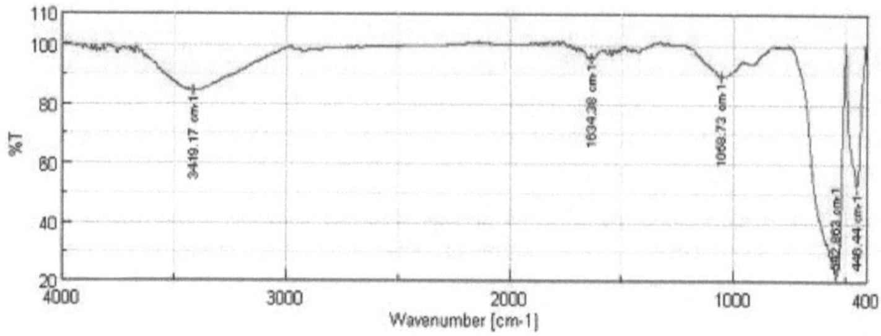

Figure 4.23: Fourier transform infrared spectrum of hematite

The hematite spectrum shows absorption bands characteristic of hematite functional groups:

- at 3419.17 cm'^{1} a weak broad band of stretching vibrations of adsorbed water molecules (v-OH) can be observed in combination with a small band of bending vibrations (δ-HrO) at 1634.38 cm"$^{(1)}$ 'corresponding to the dehydration of goethite. We also note the total disappearance of the bands located at: 1135.87 cm'(1$^{)}$(asymmetric O-band) and 1213 cm'(1$^{)}$'characteristic band at the specific oxygen surface, (γ-OH) at 796.457 and 889.987 cm"(1$^{)}$ · 604.476 cm'(1$^{)}$ valence band (γ-Fe-O), 427.155 cm'(1$^{)}$corresponding to the characteristic bands of Goethite according to Françoise Gilbert, Philippe Refait, François Leveque, Céline Remazeilles and Egle Conforto **[115].**
- a very intense band at 542.863 and a medium band at 445.44 cm'^{1} corresponding to absorption bands of the (v-Fe-O) bond in the hematite mode.

4.4.3.3 Transformation Mechanisms

The transformation of goethite (α-FeOOH) into hematite (a-PerO3) is currently a topic of great interest due to the application techniques of alpha iron oxide **[116]** as inorganic pigments, which are of great economic importance due to their chemical stability and low production costs. Iron oxides are widely used not only in paints, enamels and varnishes, but also as the main constituents of electromagnetic memory devices due to their unique magnetic properties **[117-120].**

The thermal transformation of goethite to hematite is an important step in the preparation of certain magnetic materials for recording, ferrites **[121,122]** and pigments that have been widely used due to their high colouring capacity and stability under various weather conditions: light, oxidation and corrosion **[123].** Much theoretical and experimental effort has been devoted to elucidating the transformation behaviour in such solid-state reactions.

In general, during a chemical reaction, the initial compound is destroyed and then reconstructed from the simple elements obtained to form the final compound.

This is the case for reactions occurring in an aqueous medium involving the dissolution of the initial phase followed by the precipitation of the new phase.

The crystal structures of goethite and hematite are similar, and the transition from one structure to the other can occur without any significant disruption. The atomic displacements are small and do not require a great deal of energy. In this case, the transformation can take place in the solid state at low temperatures **[124].** There are therefore simple relationships between the crystallographic axes of the two structures. The transformation is then said to be topotactic. It has been suggested that the thermal transformation reaction may be either:

(a) a direct transformation from goethite to hematite **[96,125-127]**

$$2\ \alpha\text{-}FeOOH \rightarrow \alpha\text{-}Fe_2O_3 + H_2O$$

or (b) a transformation with the creation of an intermediate phase superstructure, before the formation of hematite **[96,127].**

$$\alpha\text{-}FeOOH \rightarrow \text{a superstructure} \rightarrow Fe_2O_3$$

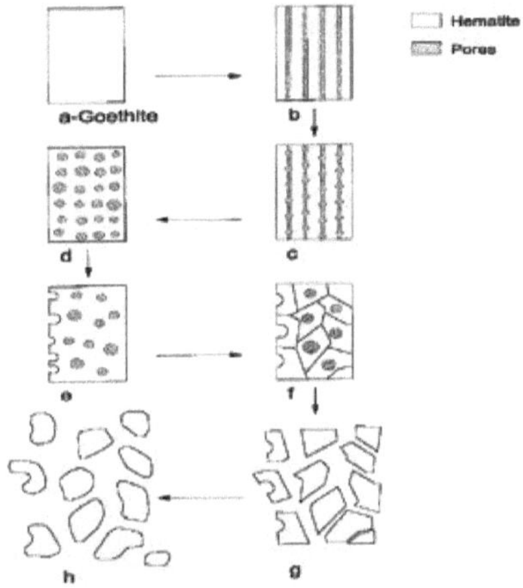

Figure 4.24: Schematic representation of the dehydration stages of goethite **[128]**

4.4.3.4 Microstructure of hematite

Haematite α-БеrОз is an iron oxide that is stable below 906°C. It is ferromagnetic up **to 768°C [129].** Its structure is rhombohedral, space group R-3c (no. 167) **[107],** constructed with iron polyhedra corresponding to highly deformed FeOe octahedra (bi-pyramidal deformation). Each trivalent iron polyhedron is connected to 13 other polyhedra by one face, three edges and six vertices; oxygen is in a tetrahedral environment of OFe4 iron. The compact stacking of the oxygen atoms is of the very slightly distorted he type. The unit cell parameters given in a hexagonal lattice are as follows: a = 5.0346Â, c = 13.752 Â.

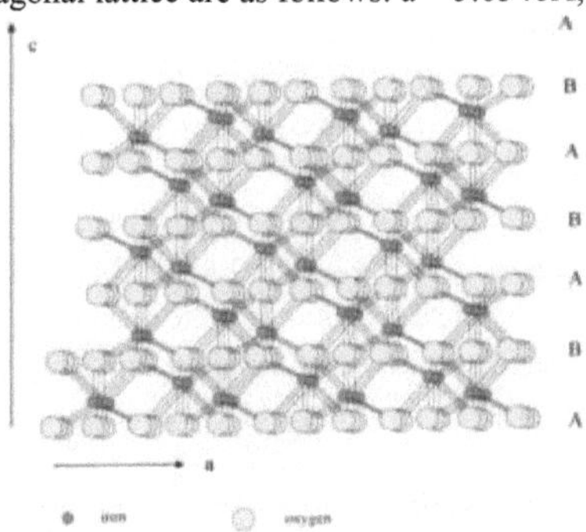

Figure 4.25: Schematic representation of the structure of hematite **[130]**

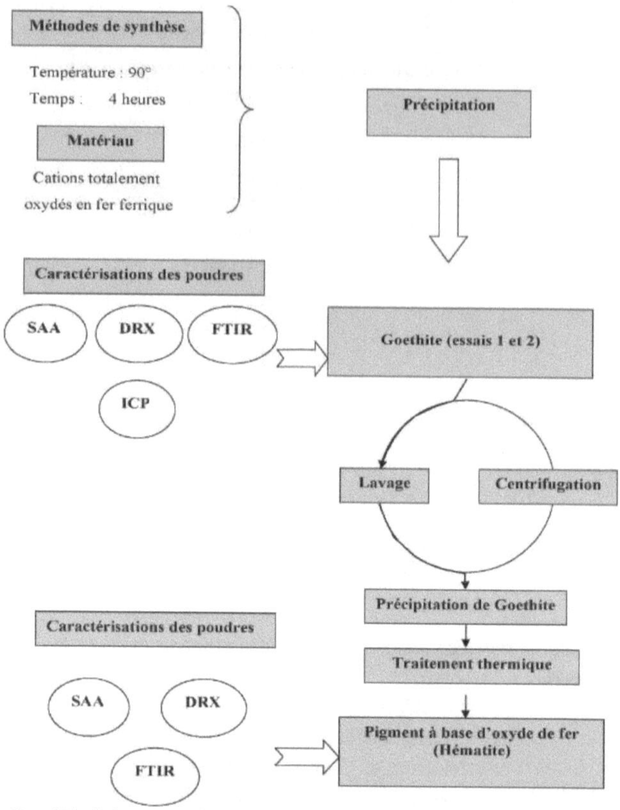

Figure 4.26: Simplified diagram of the Goethite-Hematite transformation stages

Conclusion

The work carried out as part of this project consisted of studying the recovery of zinc from solid waste from zinc hydrometallurgy. We proceeded as follows:

- Acid extraction using RC sulphuric acid solution from the electrolysis cells as the attack solution. We carried out tests at different temperatures and pH levels, modifying the reaction kinetics by varying the time.
- Extraction by double leaching followed by the Goethite process using a prepared sulphuric acid solution (return from electrolysis cells + H2SO4) as the attack solution, taking into account the physical and chemical conditions of the reactions.
- Recovery of Goethite precipitate by transforming it in a simple and economical way into an iron oxide-based pigment, which is a product of great industrial and commercial value with many uses.

Acid extraction yielded zinc recoveries of 12 to 14% with a risk of parasitic dissolution of undesirable impurities, particularly iron. The extraction of zinc in solution depends on many parameters, including the nature of the chemicals present in the solid residue, the pH of the reaction medium, the temperature and the reaction time. The optimal extraction conditions are an acidic medium (240 g/1), sufficient time for complete redissolution (120 minutes) and a temperature between 90 and 95 °C.

Extraction with Goethite precipitation following the pH interval for Goethite precipitation (test 2) yielded maximum zinc recoveries of 14.75% and 19.43%, respectively, without risk of iron and impurities dissolving. According to this method, iron is removed by forming an insoluble complex. Characterisation of the precipitate from test 1 by X-ray diffraction and FTIR (Fourier transform infrared) spectra confirmed that the compound is a mixture of goethite and hematite, with goethite being the major phase, while the precipitate from test 2 is predominantly hematite.

Tests to convert goethite into synthetic hematite, characterised by X-ray diffraction and infrared spectroscopy, revealed the complete transformation of goethite into hematite, with the goethite peaks no longer appearing in the hematite spectra.

Conclusions and prospects

The hydrometallurgical zinc extraction process involves, after roasting the blende, leaching operations aimed at solubilising the zinc in the form of ZnSO4 by dissolution in H2SO4 acid. The purified solution will yield, by electrolytic deposition, purified solid zinc marketed in the form of ingots.

During leaching, insoluble or solid cements consist mainly of complexes that are insoluble in acid leaching, with the general formula MeOPerO4, where M represents the metal ion that accompanies iron in zinc concentrates. The chemical composition varies in terms of the nature of the chemical elements, the most significant of which are iron and zinc, which can exist in several forms. Zinc ferrites are the insoluble form of zinc, which not only cause zinc losses through deposition but also accumulate in the storage of these sludges, either in the air or in technical landfills. Zinc losses have been estimated at over 21% and are a source of pollution of the natural environment.

The project in this thesis consists of optimising zinc extraction from leaching residues using acid extraction methods and iron precipitation methods in the form of a "Goethite" complex, an iron-rich compound that was then converted into hematite, an iron pigment.

Acid extraction yielded zinc recovery rates of 12 to 14% with a risk of parasitic dissolution of undesirable impurities, particularly iron. This extraction depends on many parameters, including the nature of the chemicals present in the solid residue, the pH of the reaction medium, the temperature and the reaction time. The optimal extraction conditions are an acidic medium (240 g/1), sufficient time for complete redissolution (120 minutes) and a temperature between 90 and 95°C.

Extraction by goethite precipitation was carried out at two pH intervals: 1.7>pH<2.5 and 2.5>pH<3.5.

The synthesised compounds were characterised using volumetric and gravimetric methods, atomic absorption spectroscopy and inductively coupled plasma optical emission spectrometry. The mineralogical characteristics were revealed by X-ray diffraction and Fourier transform infrared spectroscopy.

The main phases identified are goethite and hematite, with different proportions depending on the pH range chosen. The proportion of goethite is significant for pH ranges below 2.5. Goethite is dark yellow in colour with an extraction rate of 70.29%. Test precipitate 2 showed that the major phases of the hematite structure were obtained, which are red in colour with an extraction rate of 91.43%. The formation of this phase is due to the pH of the reaction medium, which promoted the transformation of goethite during the oxidation of ferrous iron to ferric iron during the treatment, while trapping only 1.82% of zinc in the iron precipitate.

The final stage of our work consists of recovering the Goethite from the second test and converting it into an iron oxide-based pigment, which is a product of great industrial and commercial value with many applications. Our choice was based on its chemical composition (Zn = 1.82% and Fe = 42.50%) and its crystallographic characteristics.

Characterisation of the compound obtained by X-ray diffraction and Fourier transform infrared spectroscopy indicated that the hematite phase was the main phase of the compound with the total disappearance of goethite. The results of the pigment obtained are in agreement with several researchers.

The solution proposed in this study is a technical solution to the problem of managing solid leachate waste, which will avoid storage in its various forms and protect the natural

environment from all forms of pollution.
The recovery of zinc from waste has been complemented by the recovery of goethite precipitates, a compound rich in iron, for use as a pigment.
We hope that this study will be continued by researching other extraction processes for other elements contained in the sludge, such as copper, cadmium, lead, etc. and to recover the precipitates obtained, such as iron-rich precipitates, for other applications such as glass manufacturing, catalysts in the coal hydro-liquefaction industry, agronomy as fertilisers, construction materials, etc., in order to neutralise these wastes and return all the unprocessed mass to nature without risk of pollution.
The results obtained in this study will open up new horizons in industrial applications instead of harmful storage of D.I.S. waste.

Bibliographical references

[1] Martine de Petris Wery and Hassine Ferid Ayedi, Electrozinc coating, engineering technique, M1600.

[2] Nicolas Seignez, doctoral thesis, University of Science and Technology of Lille I, Department of Earth Sciences, 2006.

[3] Frédéric PANFILI, doctoral thesis, University of Provence - Aix-Marseille I, 2004.

[4] J.L, Vignes, Une vie de zinc, expériences sur l'élaboration, les propriétés et le recyclage d'un matériau, Bulletin de l'union des physiciens, Voi 91, January 1997.

[5] Hsu, David D. "Chemistry: Periodic Table and More." ChemiCool. 5 May 2007. " http://www.chemicool.com "

[6] Sandrine Gombert, Air Pollution by Metals, Biosurveillance of Fallout, ADEME Collection, November 2005.

[7] Danielle Cachau-Herreillat, Experiments by the Réd-Ox family, De Boeck edition, 2006.

[8] C. R. Hammond, The Elements, wwwdO.fnal.gov/hardware/cal/lvps_info/engineering/elements.pdf, 1999.

[9] Zinc, Occurrence, uses, and properties, Encyclopedia Britannica, "www.britannica.com/EBchecked/topic/657264/zinc/8118/Occurrence-uses-and-properties ".

[10] Ronald Racek, Shaping zinc and its alloys, engineering technique, M3150.

[11] General information on zinc, ASFORMET 91, internal document from a zinc hydrometallurgy plant, ALZINC, Ghazaouet, Tlemcen.

[12] Claude Clerc, General Metallurgy Course, Thermo ARL, version 1.1, April 1990.

[13] A. Pichard, zinc and its derivatives, I NE R I S - Toxicological and environmental data sheet for chemical substances.

[14] Pierre Catilina and M-C Roure-Mariotti, Medicine and risk at work, guide for doctors in the workplace, Masson edition, 2003.

[15] Thesis, Dimitri Deneele, doctoral thesis, University of Science and Technology of Lille I, U.F.R. Earth Sciences, 2002.

[16] Laurent Rizet, Pierre-Emmanuel Charpentier, Extractive Metallurgy Hydrometallurgy, engineering technique, M-2235.

[17] Michel Darcy, Zinc Metallurgy, Engineering Technique, M-2270.

[18] C. Chaussin and G. Hilly, Metallurgy, Metal Processing, Volume 2, Dunod, Paris, 1974.

[19] Jean Barralis and Gérard Maeder, Metallurgy, processing, structures, properties, standardisation, AFNOR, Paris, 2002.

[20] Jeans Philibert, Alain Vignes, Yves Bréchet, Pierre Combrade, Metallurgy from Ore to Material, Masson edition, 1998.

[21] Fathi Habashi, A Short History of Hydrometallurgy, Hydrometallurgy, 79, 15-22, 2005.

[22] G. MILAZZO, Electrochemistry, industrial applications, Volume 2, Dunod, Paris, 1999.

[23] Fiona M. Doyle, Teaching and learning environmental hydrometallurgy, Hydrometallurgy, 79, 1 -14, 2005.

[24] Jacques Mesplède, Precis de la Chimie, cours et méthodes, Bréal edition, 2003.

[25] C. K. Gupta, T. K. Mukherjee, Hydrometallurgy in Extraction Processes, CRC Press edition, 1990.

[26] A. Benjabour, A. Benmoussat, Influence of zinc ferrites on the accumulation of solid leachate residues, Science and Technology Communication, January 2008.

[27] Pierre Blazy, El-Aid Jdid, Extractive Metallurgy, Pyrometallurgy, Engineering

Techniques, M-2231.
[28] Cüneyt Arslan, Fatma Arslan, Thermochemical Review of Jarosite and Goethite Stability Regions at 25 and 95 °C, Turkish J. Eng. Env. Sci, 27, 45- 52, 2003.
[29] William R. Richmond, Mitch Loan, Mike Newman, Gordon M. Parkinson, Zinc sulfide as a solid phase additive for improving the processing characteristics of ferrihydrite residues, Hydrometallurgy, 78,172- 179, 2005.
[30] K.R.Buban, M.J.Collins, I.M.Masters, Iron control pressure, iron and zinc control, December 1999.
[31] M.R.C. Ismael, J.M.R. Carvalho, Iron recovery from sulphate leach liquors in zinc hydrometallurgy, Minerals Engineering, 16, 31-39,2003.
[32] B. Cohen, D.S. Shipley, A.R. Tong, S.J.G. Casaroli, J.G. Petrie, Precipitation of iron from concentrated chloride solutions: Literature observations, challenges and preliminary experimental results, Minerals Engineering, 18, 1344-1347, 2005.
[33] Jussi k. Rastas, pori, Pekka J, Saikkonen, Espoo, Risto J.Honkala, Vanha-ulvila, Process of the treatment of a raw material which contains oxide and ferrite of zinc, copper and cadmium, United States patent 4355005, 19 October 1982.
[34] L. Montanaro, N. Bianchini, J.Ma. Rincon, M. Romero, Sintering behaviour of pressed red mud wastes from zinc hydrometallurgy, Ceramics International, 27, 29-37, 2001.
[35] Pappu Asokana, Mohini Saxenaa, Shyam R. Asolekar, Hazardous jarosite use in developing non-hazardous product for engineering application, Journal of Hazardous Materials, Volume 137, Issue 3, Pages 1589-1599, 11 October 2006.
[36] W. Kunda, H. Veltman, decomposition of jarosite, volume 10B, September 1979.
[37] V. Mymrin, A. Vazquez Vaamonde, Technical note: new construction materials from Spanish jarosite processing wastes, Minerals Engineering, Vol. 12, No. 11, pp. 1399-1402, 1999.
[38] Asokan Pappu, Mohini Saxena, Shyam R. Asolekar, Jarosite characteristics and its utilisation potentials, Science of The Total Environment, Volume 359, Issues 1-3, Pages 232-243, 15 April 2006.
[39] A.Van Ceulen, C. Eusebe, Iron precipitation, scientific memoirs and studies, metallurgy review, July-August, 1982.
[40] J. L. T. Hage, R. D. Schuiling, S. P. Vriend, Production of magnetite from sodium jarosite under reducing hydrothermal conditions. The reduction of Fe (III) to Fe (II) with cellulose, Canadian Metallurgical Quarterly, Vol. 38, No. 4, pp. 267-276, 1999.
[41] 39 J.O. Claassen, E.H.O. Meyerb, J. Rennieb, R.F. Sandenbergh, Iron precipitation from zinc-rich solutions: defining the Zincor Process, Hydrometallurgy, 67, 87-108, 2001.
[42] Sárka Langová, Jana Riplová, Silvie Vallová, Atmospheric leaching of steel-making wastes and the precipitation of goethite from the ferric sulphate solution, Hydrometallurgy, 87, 157-162, 2007.
[43] Alfred Richard Burkin, Chemical Hydrometallurgy, theory and principles, Imperial College Press, 2001.
[44] Amina Bendjebbour, "Extraction of zinc from solid leaching residues using the Jarosite/Alzine process", Final thesis, Abou Bakr Belkaid University - Tlemcen, 2006.
[45] P.T. Davey, T.R. Scott, Removal of iron from leach liquors by the "Goethite" process, Hydrometallurgy, 2, 25-33, 1976.
[46] Guidance document on the management of waste produced by base metal smelters, " www.ec.gc.ca/nopp/docs/rpt/mgmtWaste/fr/c5.cfm "

[47]Technical guide for zinc hydrometallurgical plants, Balen, veille montagne, Belgium.
[48]M. Romero, J.Ma. Rincon, Microstructural characterisation of a goethite waste from zinc hydrometallurgical process, Materials Letters, 31, 67-73, 1997.
[49]H. Pelino, C. Cantalino, F. Veglio, Crystallisation of glasses by recycling goethite industrial wastes to produce glass-ceramic materials, Journal of Materials Science, 29, 2087-2094, 1994.
[50]Jérôme Pradel, Simone Castillo, Jean-Pierre Traverse, Raoul Grezes-Besset, Michel Darcy, Ferric Hydroxide Oxide from the Goethite Process: Characterisation and Potential Use, Ind. Eng. Chem. Res, 32, 1801-1804, 1993.
[51]Marko Lahtinen, Leena Lehtinen, Heikki Takala, Method for precipitating iron from a zinc sulphate solution as hematite, US/0069474A1, 31 March 2005.
[52]L. Piga, L. Stoppa, R. Massidda, Recycling of industrial goethite wastes by thermal treatment, Resources, Conservation and Recycling, 14,11-20, 1995.
[53]Marcel.F.deguire, Soutthbury, John H. Wirag, Treatment of zinc calcines for zinc recovery, United States patent 4128617, 11 July 1977.
[54]M.C. Ruiz, J. Zapata, R. Padilla, Effect of variables on the quality of hematite precipitated from sulfate solutions, Hydrometallurgy, 89, 32-39, 2007.
[55]Integrated Pollution Prevention and Control (IPPC), Reference Document on Best Available Techniques in Non-Ferrous Metal Industries, December 2001.
[56]Jean Louis, Yves Torre, Process for removing iron from leaching solutions, patent Fr 2537968A1, 22 June 1984.
[57]Kaoru Saruta, Gojome-cho, Norihito Ishimori, Hitoshi Masuda, Method of leaching zinc concentrate, United States patent, US 6475450B1, 5 November 2002.
[58]L. Montanaro, N. Bianchini, J.Ma. Rincon, M. Romero, Sintering behaviour of pressed red mud wastes from zinc hydrometallurgy, Ceramics International, 27, (2001) 29-37, 2001.
[59]Roberto Orru, Mariella Sannia, Alberto Cincotti, Giacomo Cao, Treatment and recycling of zinc hydrometallurgical wastes by self-propagating reactions, Chemical Engineering Science, 54,3053-3061, 1999.
[60]E. Vahidi, F. Rashchi, D. Moradkhani, Recovery of zinc from an industrial zinc leach residue by solvent extraction using D2EHPA, Minerals Engineering, 22, (2009) 204-206, 2009.
[61]M.K. Jha, V. Kumar, R.J. Singh, Review of hydrometallurgical recovery of zinc from industrial wastes, Resources, Conservation and Recycling, 33,1-22, 2001.
[62]K.L. Bhat, K.A. Natarajan, T. Ramachandran, Electroleaching of Zinc Leach Residues, Hydrometallurgy, 18, pp. 287-303, 1987.
[63]M. Loan, O.M.G. Newman, R.M.G. Cooper, J.B. Farrow, G.M. Parkinson, Defining the Paragoethite process for iron removal in zinc hydrometallurgy, Hydrometallurgy, 81, 104129, 2006.
[64]M. Loan, T.G.St. Pierre, G.M.Parkinson, O.G.M.Newman, J.B.Farrow, Identifying nanoscale ferrihydrite in hydrometallurgical residues, JOM December 2002.
[65]Nathalie Leclerc, Eric Meux, Jean-Marie Lecuire, Hydrometallurgical extraction of zinc from zinc ferrites, Hydrometallurgy, 70, 175-183, 2003.
[66]Ichoo Sid'ahmed, Abdrrassoul Sid'ahmed, Contribution to the microbiological and physico-chem al study of seawater located near the Alzine factory, Final Year Project, Abou Bakr Belkaid University - Tlemcen 2006.
[67]Ammour Mustepha, Study of certain waste products from the ALZINC zinc plant,

Ghazaouet, 2005.
[68] S. Chergrouche, A. Bensmaili, A. Aknoun, A. Mellah, Application of the jarosite process for the removal of iron from zinc residues from the Ghazaouet electrolysis plant (Tlemcen), 17/06/2000.
[69] Y. Benarioua, M.L. Djeghlal, Contribution of zinc and silver recovery from leaching residues, June 1985.
[70] Technological guide to zinc concentrate roasting, Austriana de zinc, S.L.Metanof.
[71] M. Ziani, On-site training for production technicians in purification and leaching processes, ALZINC, 2004.
[72] M. Ziani, technical documents, On-site training for production technicians in purification processes, ALZINC, 2004.
[73] M. Ziani, technical documents, On-site training for production technicians in electrolysis processes, ALZINC, 2004.
[74] Alzine laboratory archive 2008.
[75] Practical work in analytical chemistry, "Gravimetry", www.unige.ch/sciences/chiam/williams/tp/Pharmacie/03_Gravimetrie.pdf , 2002-2003.
[76] Mendham, Denny, Barnes, Thomas, Vogel's Quantitative Chemical Analysis, Boeck edition, 2006.
[77] V. Alexéeve, Quantitative Analysis, Mir edition, Moscow, 1980.
[78] Jacques Vandegans, Anne-Marie de Kersabiec and Michel Hoenig, Atomic Absorption Spectrometry, Engineering Technique, PE 2-825.
[79] Perkin Elmer Technical Guide to Atomic Absorption, internal document from ALZINC.
[80] Francis Rouessac, Annick Rouessac, Chemical Analysis, Modern Methods and Instrumental Techniques, 3rd edition, Dunod, Paris, 2006.
[81] 8 René Poilblanc, François Crasnier, Infrared and Raman Spectroscopy, EDP Sciences, 2006.
[82] Michel Lecomte, Fourier Transform, Course and Exercises, Ecole des Mines de Douai, July 2001.
[83] J.Michel Hollas, Spectroscopy, Dunod, Paris, 2003.
[84] Thomas Belin, doctoral thesis, University of Burgundy - UFR Sciences et Techniques, Ecole Doctorale Carnot, 2002.
[85] Lesley Smart, Elaine Moore, Introduction to Solid State Chemistry, Course and Exercises with Answers, Masson, 1997.
[86] Norbert Broil, Characterisation of crystallised solids by X-ray diffraction, engineering techniques, PE 1-080.
[87] Graziella Marc, X-ray diffraction, "aasaa.free.fr/Methodes/Dx/DiffractionX".
[88] Agnès Hagège, Alain Lamotte, Maurice Leroy, Analysis of traces and ultra-traces of elements, engineering techniques, P 3-860.
[89] Jean-Michel Mermet, Emmanuelle Poussel, High-frequency induced plasma coupling - mass spectrometry, engineering techniques, P 2 720.
[90] Morgan Barale, doctoral thesis, University of Paris 6, 2006.
[91] Société des mines et fonderies de zinc de la veille montagne, Process for recovering zinc-s from zinc-bearing material leaching residues, patents BE 724214, 1972.
[92] Christelle Not, doctoral thesis, University of Quebec in Montreal, March 2006.
[93] Société des mines et fonderies de zinc de la veille montagne, Process for preparing an

iron oxide-based pigment, patents FR 2.135.141, 15-12-1972.
[94] Presentation of the zinc ingot production plant using hydrometallurgy
ALZINC Ghazaouet - Tlemcen (internal document).
[95] Fatiha. Mostefa Kara, Mohamed. Hadjel, Abdelaziz Bendraoua, Elimination of iron present in effluents from the AL-ZINC zinc electrolysis unit in GHAZAOUET, 4 June 2008.
[96] M. Pelino, Recycling of zinc-hydrometallurgy wastes in glass and glass ceramic materials, Waste Management, 20, 561-568, 2000.
[97] Loie Francke, doctoral thesis, University of Chemical Sciences, University of Bordeaux 1, 16 October 2002.
[98] H. D. Rúan, R. L. Frost, J. T. Kloprogge and L. Duong, Infrared spectroscopy of goethite dehydroxylation. II. Effect of aluminium substitution on the behaviour of hydroxyl units, Spectrochimica Acta Part A: Molecular and Biomolecular Spectroscopy, Volume 58, Issue 3, 479-491, February 2002.
[99] Cloudio Morterra, Anna Chiorina, Enzo Borrello, An IR Spectroscopic Characterisation of a- FeOOH (Goethite), Materials Chemistry and Physics, 10, 119-138, 1984.
[100] H.D. Rúan, R.L. Frost, J.T. Kloprogge, L. Duong, Infrared spectroscopy of goethite dehydroxylation: III. FT-IR microscopy of in situ study of the thermal transformation of goethite to hematite, Spectrochimica Acta Part A, 58, 967-981, 2002.
[101] Jean-Francois Body, Andrew R. Felmy, On the protonation of oxo- and hydroxo-groups of the goethite (a-FeOOH) surface: A FTIR spectroscopic investigation of surface O-H stretching vibrations, Geochimica et Cosmochimica Acta 72, 3338-3357, 2008.
[102] C.A. Luiz Oliveira, C. Teodorico Ramalho, F.Eugênio Souza, Maraísa Gonçalves, Q.L. Diana Oliveira, C.Márcio Pereira, D.José Fabris, Catalytic properties of goethite prepared in the presence of Nb on oxidation reactions in water: Computational and experimental studies applied Catalysis B: Environmental, Volume 83, Issues 3-4, 169-176, 23 September 2008.
[103] H. Namduri, S. Nasrazadani, Quantitative analysis of iron oxides using Fourier transform infrared spectrophotometry, Corrosion Science, 50, 2493-2497, 2008
[104] 104] L. Verdonck, S. Hoste, F.F. Roelandt, G.P. Van Der Kelen, Normal coordinate analysis of a-FeOOH - a molecular approach, Journal of Molecular Structure, 79, 273-279, 1982.
[105] P.S.R. Prasada, K. Shiva Prasada, V. Krishna Chaitanya, E.V.S.S.K. Babu, B. Sreedhar, S. Ramana Murthy, In situ FTIR study on the dehydration of natural goethite, Journal of Asian Earth Sciences 27, 503-511, 2006.
[106] Russell, J.D., Fraser, A.R., In: Infrared Methods in Clay Mineralogy: Spectroscopic and Chemical Determinative Methods, Wilson, MJ (Ed.), Chapman & Hall, London, 49-52, 1994.
[107] R. M. Cornell, Udo Schwertmann, Iron oxides book, Structure, Properties, Reactions, Occurrences and Uses, copyrighted material, 2nd edition 2003.
[108] Nora H. de Leeuw a, b, Timothy G. Cooper b, Surface simulation studies of the hydration of white rust Fe(OH)2, goethite a-FeO(OH) and hematite a-EerO3, Geochimica et Cosmochimica, Acta 71, 1655-1673, 2007.
[109] C. Betiana Campo, B. Olivier Rosseler, C. Mariana Alvarez, H.C Elsa Rueda, A. Maria Volpe, On the nature of goethite, μη-goethite and Co-goethite as supports for gold- , Materials Chemistry and Physics, 109, 448-454, 2008.
[110] Yannick Cudennec, André Lecerf, Topotactic transformations of goethite and

lepidocrocite into hematite and maghemite, Solid State Sciences, 7,520-529, 2005.

[111] A. B Daniel Tunega, A. H Martin Gerzabek, C.Georg Haberhauer, U Kai. Totsched, B. Hans Lischka, Model study on sorption of polycyclic aromatic hydrocarbons to goethite, Journal of Colloid and Interface Science 330, 244-249, 2009.

[112] Thomas Belin, Doctoral thesis, University of Burgundy, 27 November 2002.

[113] P. Segalen, Iron in soils, O. R. S.T.O.M,PAR I S, 1964.

[114] Thomas Verdier, Thomas Lebrun, Virginie Nivoix, Malick Jean, Béatrice Hannoyer, Synthesis of mixed iron and zinc oxides by high-energy grinding, 2002.

[115] Françoise Gilbert, Philippe Refait, Francois Leveque, Celine Remazeilles, Egle Conforto, Synthesis of goethite from Fe(0H)2 precipitates: Influence of Fe(II) concentration and stirring speed, Journal of Physics and Chemistry of Solids, 69, 2124-2130, 2008.

[116] L. Diamandescu, D. MihBilA TarBbAganu, On the solid phase transformation goethite, hematite, Materials Letters, 17, 309-311, 1993.

[117] Honglei Fan, Baozhen Song, Qiaoxia Li, Thermal behaviour of goethite during transformation to hematite, Materials Chemistry and Physics, 98, 148-153, 2006.

[118] A.B Xiaohe Liu, B. Guanzhou Qiu, B. Aiguo Yan, A. Zhong Wang, A.Xingguo Li, Hydrothermal synthesis and characterisation of a-FeOOH and a-PerO3 uniform nanocrystallines, Journal of Alloys and Compounds, 433,216-220, 2007.

[119] Ming-Wang Shao, Hui-Zhao Ban, Yan-Hua Tong, Hui Hu, Li-Ling Niu, Hua-Zhong Gao, Yin Ye, Synthesis of single-crystalline alpha-FeOOH nanorods with semi-hard magnetic property, Materials Letters, 61, 4318-4320, 2007.

[120] Han Yaoxing, Ma Xinsheng, Cao Hongming, Zhang Haiying, Wu Qiufang, Morphological study and thermal analysis of surface modified a-FeOOH via in situ polymerisation of methyl methacrylate, Materials Research Bulletin, 39, 1159-1166, 2004.

[121] A.B. M. Uda, A.S. Sassa, C.S. Yoshimura, D.J. Kondo, E.M. Nakamura e, A.Y. Ban, F.H. Adachi, Yellow, red and blue pigments from ancient Egyptian palace painted walls, Nuclear Instruments and Methods in Physics Research B, 161-163, 758-761, 2000.

[122] P.F. Rossi, Heats of immersion of iron oxide pigments in water: Goethite and Hematite, Materials Chemistry, 3,7- 16, 1978.

[123] D. Bikiaris, Sister Daniilia, S. Sotiropoulou, O. Katsimbiri, E. Pavlidou, A.P. Moutsatsou, Y. Chryssoulakis, Ochre differentiation through micro-Raman and micro-FTIR spectroscopies: application on wall paintings at Meteora and Mount Athos, Greece, Spectrochimica Acta Part A, 56, 3-18, 1999.

[124] J. Murray, L. Kirwan, M. Loan, B.K. Hodnett, In-situ synchrotron diffraction study of the hydrothermal transformation of goethite to hematite in sodium aluminate solutions, Hydrometallurgy, 2008.

[125] Fumio Watari, J.M. Cowley, The study of oxide formation on (001), (Oil), (111) and (113) surfaces of Cr thin films using stem-micro-diffraction methods surface Science, Volume 105, Issue 1, 240-264, 1 April 1981.

[126] Fumio Watari, P. Delavignette, J. Van Landuty, S. Amelinckx, Electron microscopic study of dehydration transformations. Part III: High resolution observation of the reaction process FeOOH ΓerO3, Journal of Solid State Chemistry, Volume 48, Issue 1, 49-64, June 1983.

[127] Fumio Watari, P. Delavignette, S. Amelinckx, Electron microscopic study of dehydration transformations. II. The formation of "superstructures" on the dehydration of goethite and diaspore, Journal of Solid State Chemistry, Volume 29, Issue 3, 417-427,

September 1979.
[128] M. PomieÁ, M. Menu, C. Vignaud, TEM Observations of Goethite Dehydration: Application to Archaeological Samples, Journal of the European Ceramic Society, 19, 1605-1614, 1999.
[129] Moussa Grafoute, doctoral thesis, University of Maine, 7 October 2004.
[130] Yannick Cudennec, Andre Lecerf, The transformation of ferrihydrite into goethite or hematite, revisited, Journal of Solid State Chemistry, 179, 716-722, 2006.

Appendix A
Chemical dosages

- **Calcium dosages**

From the iron filtrate obtained when the leaching residues are dissolved, boil, add 15 cc of NH4OH, 25 cc (M l.ihtfO.i ammonium oxalate, boil for a few minutes, then leave to stand on an asbestos plate, filter through a hot blue filter, rinse the filter with hot water, divide the filter in two and place in the Erlenmeyer flask, add 20 cc H2SO4 1/1, 200 cc of hot water at 70°C. Titrate with KMNO4, colour change from colourless to pink.

Ca %= V*T

- **Magnesium dosage**

From the hot calcium filter, stir for 15 minutes, add 50 cc $(NH_4)_2 HPO_4$, stir for 15 min, add NH4OH until the solution becomes clear, leave to stand overnight, filter through a cold blue filter, dry the filter in an oven and calcine at 1000°C, weigh the crucible before and after calcination.

Mg %= (A-B) * 21.85

- **Total sulphur content**

Weigh 0.5 g of blende in a 500 cc Erlenmeyer flask, add 10 cc H2O, 10 cc NaClOs, 15 cc HNO3, bring to see, add 25 cc HCl, bring to see, add 10 cc HC1, 50 cc H2O, boil and filter through a red filter, wash the filter with boiling water then boil the filtrate, add 30 cc BaCh, boil and leave to rest on an asbestos plate, filter through a blue filter, dry and calcine.

S %= (A-B) * 13.74*2

Appendix B

Characteristics of the various phases encountered in this work

8.1 Hematite a-TerO3

Iron oxide (III) crystallising in the hexagonal system (space group A 3cyMedia parameters: a= 0.50356 nm and b= c= 1.37489 nm **[1]**.

Table B.l: Indexes, positions and relative intensities of the diffraction lines of hematite a-TerO3 (incident wavelength λ Cu= 0.15406 nm and dHKL given in nm) [extracts from JCPDS file 33-0664].

hkl	20	dHKL	I=Io	hkl	20	dHKL	I=Io	hkl	20	dHKL	I=Io
012	24.138	0.368	30	11	54.091	0.169	45	1010	71,937	0.131	10
104	33.153	0.270	10	211	56.152	0.164	1	119	*72.262*	0.130	6
110	*35,612*	0.252	70	122	57,429	0.160	5	220	75.43	0.126	8
006	39.277	0.229	3	018	57.59	0.159	10	306	77.73	0.123	4
113	40.855	0.221	20	214	62.451	0.148	30	223	78.76	0.121	2
202	43.519	0.208	3	300	*63,991*	0.145	30	128	80,711	0.119	5
024	49.480	0.184	40	208	69,601	0.135	3	0210	82.94	0.116	5

Table B.2: Indexations, positions and relative intensities of the diffraction lines of hematite a-TerO3 (incident wavelength λ Co= 0.1789 nm and dHKL given in nm) [extracts from JCPDS file 33-0664].

hkl	2	dHKL	I=Io	hkl	20	dHKL	I=Io	hkl	20	dHKL	I=Io
012	28.104	0.368	30	11	63.741	0.169	45	1010	86.005	0.131	10
104	38.694	0.270	10	211	66.257	0.164	1	119	86.423	0.130	6
110	41.599	0.252	70	122	67.822	0.160	5	220	90.528	0.126	8
006	45.942	0.229	3	018	68.019	0.159	10	306	93.546	0.123	4
113	47.819	0.221	20	214	74.024	0.148	30	223	94.910	0.121	2
202	50.995	0.208	3	300	75.944	0.145	30	128	97.514	0.119	5
024	58.153	0.184	40	208	83.017	0.135	3	0210	100.526	0.116	5

8.2 Goethite α-FeOOH

Iron hydroxide crystallising in the orthorhombic system. Madie parameters: a= 9.95 Å, b= 3.01 Å and c= 4.620 Å **[1]**.

Table B.3: Indexes, positions and relative intensities of the diffraction lines of a-FeOOH hematite (incident wavelength λ Cu= 0.15406 nm and dHKL given in nm) [extracts from JCPDS file 29-0713].

hkl	20	dHKL	I=Io	hkl	20	dHKL	I=Io	hkl	20	dHKL	I=Io
20	17.78	0.498	12	401	41.16	0.219	18		59.08	0.156	8
101	21.22	0.418	100	311	45.04	0.201	2		61.36	0.151	8
201	26.32	0.338	10	410	47.28	0.192	5		63.3	0.146	2
301	33.24	0.269	35	112	50.6	0.18	6		63.96	0.145	5
210	34.70	0.258	12	211	53.24	0.171	20		65.64	0.142	2
111	36.64	0.244	50		54.2	0.169	6		68.44	0.136	2
211	39.96	0.225	14	-	50	0.156	10		71.54	0.131	3

8.3 Franklinite

Zinc ferrites with the general formula (Zn,Mn,Fe)(Fe,Mn)2O4, crystallising in the cubic system ( ). Lattice parameters: a= 8.466 Â **[1].**

Table B.4: Indexations, positions and relative intensities of the diffraction lines of zinc ferrites (incident wavelength λ Cu= 1.5406 nm and dHKL given in nm) [1].

20	dHKL	I=Io
29.8	2.990	70
35.16	2.55	100
42.61	2.120	40
52.87	1.73	30
56.32	1.632	70
61.84	1.499	80
88.49	1.104	50
103.92	0.978	50

Tableau 5 5: Indexations, positions and relative intensities of zinc ferrite diffraction lines (incident wavelength λ Cu= 1.789 nm and dHKL given in nm)

2	dHKL	I=Io
3	2.99	70
41.07	2.55	100
49.91	2.120	40
62.26	1.73	30
66.47	1.632	70
73.27	1.499	80
108.23	1,104	50
123	0.978	50

B.4 Anglesite

Lead sulphate with the general formula PbSO4, crystallising in the orthorhombic system.
Media parameters: a= 6.959 Â, b= 8.482 Â, c= 5.398 Â **[1].**

Tableau 6 6: Indexes, positions and relative intensities of Anglesite diffraction lines (incident wavelength λ Cu= 1.5406 nm and dHKL given in nm) [1].

20	dHKL	I=Io
20.8	4.2	90
23.32	3.810	60
26.74	3.330	90
27.68	3.220	70
29.75	3	100
33,165	2,699	50
43.76	2,067	80
44.64	2.028	50

B.5 Sphalerite

Zinc sulphide with the general formula ZnS, crystallising in the cubic system.
Madie parameters: a= 5.409 Â **[1].**

Tableau 7 7: Indexations, positions and relative intensities of Anglesite diffraction lines (incident wavelength λ Cu= 1.789 nm and dHKL given in nm) [1].

20	dHKL	I=Io

3	3.12	10
38.62	2.705	10
55.78	1.912	50
66.42	1.633	30
82.92	1.351	10
92.33	1.24	10
108.33	1.103	10

Tableau 8 8: Indexations, positions and relative intensities of Anglesite diffraction lines (incident wavelength λ Cu= 1.5406 nm and dHKL given in nm) [1].

2θ	dHKL	I=Io
28.55	3.123	10
33.09	2.705	10
47.51	1,912	50
56.29	1.633	30
69.52	1.351	10
76.8	1.24	10
88.59	1.103	10

B.6 Zincite

Zinc oxide with the general formula (Zn, Mn) O, crystallising in the hexagonal system. Lattice parameters: a= 3.25 Å, c= 5.207 Å.

Table B.9: Indexations, positions and relative intensities of Anglesite diffraction lines (incident wavelength λ Cu= 1.789 nm and dHKL given in nm) [1].

2θ	dHKL	I=Io
37.0	2.81	70
40.21	2.602	60
42.35	2,476	100
55.81	1.911	30
66.75	1.626	40
74.54	1.477	30
80.88	1.379	30
82.32	1.359	10

Table B.10: Indexations, positions and relative intensities of Anglesite diffraction lines (incident wavelength λ Cu= 1.5406 nm and dHKL given in nm) [1].

2θ	dHKL	I=Io
31.7	2.81	70
34.35	2.602	60
36.25	2,476	100
47.54	1.911	30
56.55	1.626	40
62.86	1.477	30
67.91	1.379	30
69.05	1.359	10

Appendix C
Spectrum of goethite and hematite

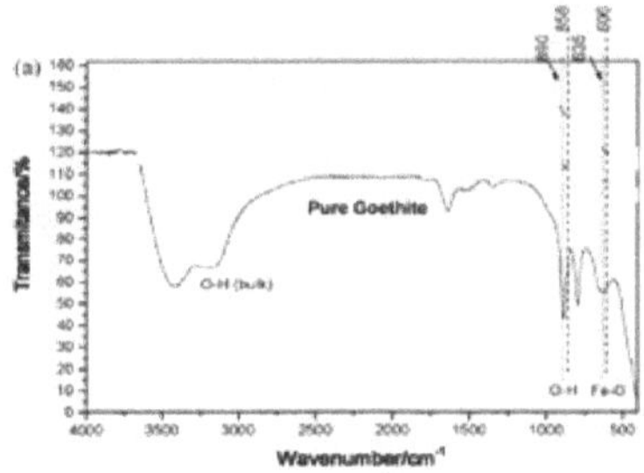

Figure C.1: FTIR spectrum of pure goethite **[3]**

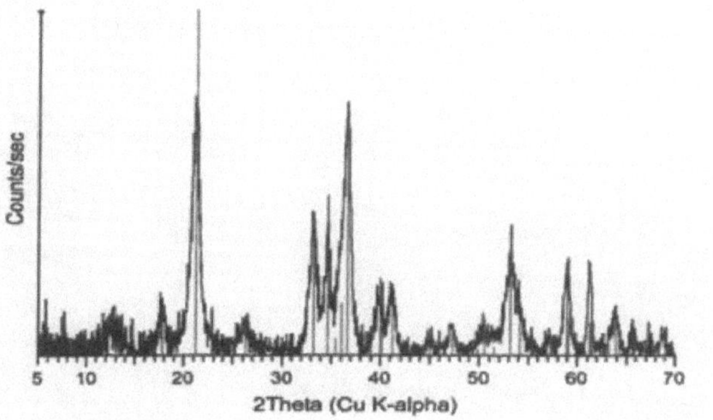

Figure C.2: X-ray diffractogram of a hematite pigment obtained by calcination of goethite
(precipitated from iron waste at 750°C [2]

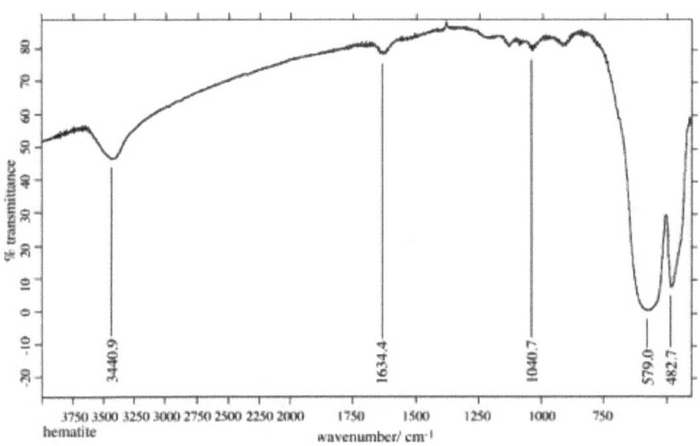

Figure C.3: FTIR spectrum of hematite pigment [4]

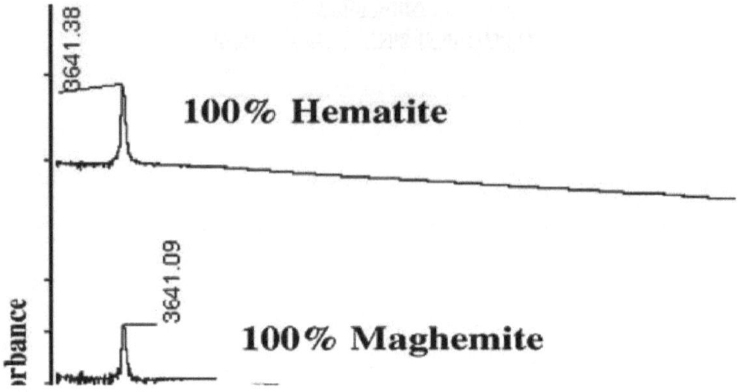

Figure C.4: FTIR spectrum of pure hematite [5]

Appendix bibliography

[1] Mineralogical Atlas, "http://webmineral.brgm.fr:8003/mineraux/Main.html" (Iron and iron compounds), Paris Museum collection, photograph by U. de Cayeux, BRGM, May 1997.

[2] M.A. Legodi, D. de Waal, The preparation of magnetite, goethite, hematite and maghemite of pigment quality from mill scale iron waste, Dyes and Pigments 74 161-168, 2007.

[3] Luiz C.A. Oliveira, Teodorico C. Ramalho, Eugênio F. Souza, Maraísa Gonçalves, Diana Q.L. Oliveira, Márcio C. Pereira, José D. Fabris, Catalytic properties of goethite prepared in the presence of Nb on oxidation reactions in water: Computational and experimental studies, Applied Catalysis B: Environmental, Volume 83, Issues 3-4, 169-176, 23 September 2008.

[4] DirkWalter, Characterisation of synthetic hydrous hematite pigments, Thermochimica Acta 445, 195-199, 2006.

[5] **H.** Namduri, **S.** Nasrazadani, Quantitative analysis of iron oxides using Fourier transform infrared spectrophotometry, Corrosion Science, 50, 2493-2497, 2008.

FSC
www.fsc.org
MIX
Papier | Fördert gute Waldnutzung
FSC® C083411

Kolibri 360 GmbH
Ferdinand-Jühlke-Straße 7
99095 Erfurt, Deutschland
produktsicherheit@kolibri360.de

Druck:
CPI Druckdienstleistungen GmbH
im Auftrag der
Kolibri 360 GmbH
Ferdinand-Jühlke-Str. 7
99095 Erfurt